OBLIVION

AMERICA AT THE BRINK

---

by

Lt. Col. Thomas E. Bearden
(U.S. Army Retired)

This book is an expanded version of a very close-hold brief provided to a certain Head of State and his Foreign Minister in 2003.

OBLIVION: America at the Brink

First Edition 2005

Cheniere Press
Santa Barbara, California 93109
USA

2

www.cheniere.org

ISBN: 0-9725146-2-7

Dedicated to my beloved wife, Doris, without whose patience and support none of this would have been possible.

Table of Contents

Strategic Aspects of the Asymmetric War on the U.S.

Slide Briefing

Strategic Aspects of the Asymmetric War on the U.S.

Including Scalar Electromagnetic Weapons and their Terrorist Uses

T. E. Bearden

First edition Oct. 13, 2004

Drafted Sept. 11, 2004 on the Third Anniversary of 9/11/01

Revised and updated 23 April 2005

Cold War Development of Scalar Interferometry

Prior to 1990, the weather over North America was being steadily engineered by the KGB — particularly beginning on July 4, 1976 — using giant strategic scalar interferometers on site in Russia {i, ii, iii, iv}. From the beginning of the development program which started shortly after WW II, the KGB personally controlled the development of startling new Russian weapons under the Soviet *energetics* program {v, vi, vii}. That program was for the research and eventual development of highly advanced new superweapons more powerful than the atomic bomb.

The new superweapons were developed, produced, manned, and operated by the KGB itself, and were never placed in the hands of the regular Russian armed forces. Speaking to the Presidium in 1960, Khrushchev referred to the forthcoming scalar interferometer weapons with the following statement:

> "We have a new weapon — just within the portfolio of our scientists, so to speak — so powerful that, if unrestrainedly use, it could wipe out all life on earth."

The large Soviet strategic scalar interferometers were deployed and became operational in April 1963 — a bit too late for use by Khrushchev to counter the U.S. confrontation in the Cuban Missile Crisis of latter 1962. However, the first operational weapon was used to destroy the U.S.S. Thresher nuclear submarine while on underwater maneuvers off the East Coast of the United States, in April 1963. Thereby Khrushchev demonstrated the power of his new weapons over nuclear submarines — one of the major three elements of the strategic military nuclear firepower of the United States. Extension to the other two elements — intercontinental ballistic missiles (ICBMs) and strategic bombers — was obvious.

Incidents Where Outside Intervention Saved the United States

Many dramatic but largely unheralded incidents occurred during the so-called "Cold War". For example, in the 1970s a small nation saved the United States from total destruction by Soviet nuclear weapons which had been inserted in our cities. That nation did it very simply: they likewise inserted an appreciable number of nuclear weapons of their own, some thermonuclear, inside Russia itself, in its population centers and its primary target zones.

All these inserted nuclear weapons — both in the United States and in Russia — are still there with their activation teams, waiting. In such manner was the "balance of terror" — the Mutual Assured Destruction or MAD doctrine — really implemented and enforced vis à vis the Soviet Union. At the time this approach to deterrence was called "Dead Man Fuzing".

In April 1986 a private little U.S. group intervened to prevent a forthcoming giant earthquake that was being built up for the greater Los Angeles and San Francisco areas by the KGB scalar interferometers. Via a special electronic device, the group suddenly destroyed one of the distant Soviet scalar interferometer transmitters — thereby initiating the nearby Chernobyl nuclear incident but preventing the loss of perhaps 200,000 U.S. lives in Los Angeles and San Francisco, together with preventing terrible destruction and economic damage.

Also in 1986, the same friendly little nation again prevented our strategic destruction by even more powerful energetics weapons developed by the Soviet Union. As early as 1984, the friendly nation simply began exploding very large Russian missile ammunition storage sites, to warn the Soviet Union of what was in store for it, if it attacked.

In 1997, again the U.S. would have been utterly destroyed on two occasions had it not been for the same little nation. Several professional colleagues and I played a desperate role in both of those frighteningly real incidents, which were never reported in the news media. Indeed, our own intelligence agencies had no inkling that a full strategic Soviet superweapon attack to destroy the United States was imminent on each of the occasions (the second of which would have occurred on May 1, 1997) {viii}. The May Day attack for 1997 would have been the "big one", had the Federal Security Service (FSB/KGB) not been checkmated by "an offer it could not refuse" — delivered once again by the friendly nation. During those critical periods when every hour on the clock seemed a week in length, the hostile armada that would have struck us included the Japanese Yakuza and Aum Shinrikyo teams on site in Russia, operating leased KGB/FSB strategic scalar interferometers against the U.S.

Yakuza Acquisition of Soviet Superweapons

With the collapse of the old Soviet Union's economy, by the end of 1989 the Russian financial situation was grim. By courtesy of the Russian Mafia {ix} — which works directly for the FSB/KGB — a rogue Japanese group consisting of the Yakuza {x} and the Aum Shinrikyo cult {xi} was conducted to Russia in latter 1989, met with the KGB, and leased some of those large KGB strategic scalar interferometers on site in Russia {xii}. The down payment was $900 million U.S. in gold bullion, and the lease is rumored to be some $1 billion per year. The payment was no problem for the Japanese team. The Yakuza had ripped off the

Japanese taxpayer for hundreds of billions of dollars, never repaid, and so the rogue Japanese group could easily afford such leasing {xiii}.

The Yakuza and Aum Shinrikyo then assigned teams to man those superweapon sites and begin weather engineering operations against the United States and other targets. The Aum Shinrikyo even set up a small university in Russia where the energetics theory and technology could be taught to the rogue Japanese teams by the KGB scientists.

After much training, in 1990 the rogue Japanese teams began operating the leased interferometers {xiv} worldwide for weather engineering and other purposes under the supervision of the FSB/KGB. The Yakuza and its vast resources worldwide were thus incorporated into the FSB/KGB planning and coordination for the coming destruction of the United States and the Western world.

Much later, Western investigations into the Aum Shinrikyo did clearly establish the sect's links in Russia, but missed the major involvement of the Yakuza. The Soviets easily made it appear that contact of the Aum Shinrikyo with Russia had been harmless (with the exception of the sarin nerve gas). Quoting Turbiville {xv}:

> "Russian, Japanese, and other investigators quickly identified a substantial number of [Aum Shinrikyo] sect members in Russia, including members in the government and other walks of life. Media commentators took note of a 'gas analyzer' of Russian origin seized at a sect facility; reports of Aum Shinrikyo sect members among Russian Radiation, Chemical, and Biological Defenses Troops; alleged sect ties to the Russian Academy of Sciences; the large volume of commercial Russian ship and aircraft traffic between Russia and Japan; and other issues that suggested questionable Russo-Japanese linkages to sarin production or transport... Official Russian military and security service spokesmen, while acknowledging Aum Shinrikyo's presence in Russia, reiterated the absolute security of military chemical depots and munitions..."

So it was that in early 1990, the FSB/KGB transferred their North American weather engineering operation to the Yakuza/Aum Shinrikyo teams, who perpetuated it using the Russian scalar interferometers that the Yakuza had leased. Today the rogue Japanese teams on site in Russia and elsewhere continue their weather engineering against the United States under FSB/KGB supervision.

At the end of 2004, the FSB/KGB and Yakuza entered the 2-year "Operations Phase" of asymmetric war against the United States. Weather operations — including Hurricanes Charlie, Frances, Ivan, Jeanne, etc. — intensified and remain intense, wreaking great economic damage. The program included directly influencing and controlling each hurricane's power and behavior, as well as directing its course, speed and targeting path. Indeed, Ivan did a 180 degree turn, and Jeanne did a 360 degree loop before reaching Florida, demonstrating an unprecedented and sophisticated level of storm control.

Meteorologists do recognize periods of increased or decreased hurricane activity for various reasons {xvi}, but very few consider deliberate human induction of hurricanes or human control over their direction, power, and progress {xvii}. Fortunately, a few do.

Indeed, in latter March of 2004, Hurricane Catarina — the *first-ever* recorded in the South Atlantic — formed and came ashore in Brazil on March 28 with 90 mph winds, doing substantial damage. So while the conventional wisdom is that hurricanes cannot form (naturally) in the South Atlantic, this one did and "broke all the records". It appears to have been a deliberate probe by the Yakuza: *Produce and drive ashore a hurricane where the textbooks state one is impossible, to test whether Western governments and scientists recognize the artificial weather engineering*. The answer, of course, is that — as expected — the West did not recognize its importance, or that it was a deliberate "stimulus." Western meteorologists and governments simply shrugged off Hurricane Catarina as an interesting little phenomenon of no great concern.

Some time after their 1989 lease of the Russian interferometers, the Yakuza carried the science and technology of scalar electromagnetics (energetics) back to Japan. There they set up their own clandestine facilities to manufacture such weapons, including small portable scalar interferometers, electromagnetic pulse (EMP) weapons, and even *negative energy* EMP weapons. Any group or nation possessing scalar interferometers is also able to produce either positive energy EMP or negative energy EMP in the distant interference zone. Smaller, more portable weapons then follow.

Selected portable weapons of such types will be inserted — probably have already *been* inserted — into the U.S. for destroying its key internal targets and centralized electrical power system. With the final *coup de grace* to be delivered about two years from now, the loss of our national electrical power system is intended to evoke the catastrophic collapse of the entire U.S. economy, followed by the fall of other Western nations' economies like toppling dominoes.

Despite the risk, the U.S. Government, scientific community, intelligence community, and electrical power industry seem totally incapable of confronting the desperate requirement to replace our entire centralized electrical power system with fuel-free self-powering "energy from the vacuum" (EFTV) systems as rapidly as possible. Merely mentioning the requirement evokes psychological displacement activity, denial, and fierce personal attacks on the proponent. It also induces strong opposition from the powerful cartels in our society — energy, science, government, the environmental community, the financial and industrial community, etc. Hence the U.S. continues "energy business as usual", while the clock ticks away to our destruction {xviii}.

Our Desperate Need to Correct the Seriously Flawed Classical Electromagnetics (CEM) and Electrical Engineering (EE) Models

In the past we have described the longitudinal electromagnetic (EM) wave transmitters used in these interferometer weapons with the phrases "scalar electromagnetics" and "scalar interferometry" {xiv}. Tesla did not originate these terms, nor did he originate scalar interferometry, although he had indeed stumbled onto longitudinal EM waves and what can only be called *force-free precursor engineering*_{xix}. Quoting Tesla {xx}:

> "...I showed that the universal medium is a gaseous body in which only longitudinal pulses can be propagated, involving alternating compressions and expansions similar to those produced by sound waves in the air. Thus, a wireless transmitter does not emit Hertz waves which are a myth, but sound waves in the ether, behaving in every respect like those in the air, except that, owing to the great elastic force and extremely small density of the medium, their speed is that of light."

If "gaseous body" is replaced with the modern term "virtual particle flux (active *virtual state* gas) of the vacuum", Tesla's words agree with the basic view of the modern active vacuum.

Whittaker's 1903-4 discovery that sets of longitudinal EM waves actually *comprise* all normal EM waves, fields and potentials {xxi,xxii} leads to a much more fundamental electrodynamics in which ordinary EM waves, potentials, and fields can undergo sophisticated alteration, enabling them to contain hidden internal field vectors and dynamics. ("Dynamics" refers to change over time.)

Indeed, the so-called "transverse EM waves" measured by our instruments are merely precession wave envelopes of charged particles (charged masses) responding gyroscopically at right angles to the more fundamental longitudinal EM waves. Longitudinal waves are force-free except when interacting with matter.

Transverse EM force field and force field waves exist only in association with charged matter, never in free space. What exist in space as the "precursor field" are the Whittaker sets of bidirectional longitudinal EM wave pairs, for each force field wave, force field, or potential in charged matter when the precursor wave sets are interacting with that charged matter. What Whittaker showed as the "decomposition" of such waves is actually the precursor longitudinal force-free EM wave *mixture* existing in charge-free space, whose summation — via the wave interaction with charged matter — yields transverse force field waves. An EM wave in space (i.e. where no matter is present) actually consists of a dense, multi-frequency mixture of bidirectional longitudinal EM wave pairs. It *is not* a "pale thin shadow" of the form of the matter wave in charged matter.

Every EM "field" and "wave" and "potential" in space — including so-called "static" fields and potentials — actually exists as a large ensemble of ongoing EM energy flows in the form of those propagating longitudinal EM wave sets. But our standard classical EM and electrical engineering model's false assumption of a material ether has prevented recognition of the fundamental character of these longitudinal EM wave ensembles. The model continues to erroneously assume *forces and force fields* in space. Yet forces and force fields cannot occur in space but only in mass, because one of the components of force is mass, by the definition $\mathbf{F} \equiv d\mathbf{p}/dt = d/dt(m\mathbf{v})$. Simply replace m by 0, and $\mathbf{F} = 0$.

When electrons are in a conductor, their travel is limited to a few inches per second down the wire in a typical nominal case. Thus, when an electron in a wire experiences a longitudinal EMF, it has to precess laterally. Most of its motion is perpendicular to the wire. This rotational movement is what generates the transverse EM waves that our instruments measure, as the electrons respond to the more fundamental longitudinal vibrations. At least Jackson {xxiii} clearly states what CEM/EE assumes:

> "Most classical electrodynamicists continue to adhere to the notion that the EM force field exists as such in the vacuum, but do admit that physically measurable quantities such as force somehow involve the product of charge and field."

So the *measured* EM transverse force **F** is the effect of a more primal, precursor, force-free EM field (longitudinal precursor waves) in space, *in and on* charges q in the ongoing interaction of that precursor EM field with those charges q. To see this, examine the simple equation $\mathbf{F} = \mathbf{E}q$, as a definition $\mathbf{F} \equiv \mathbf{E}q$. In the absence of charged matter, $q = 0$ and $\mathbf{F} = 0$, but **E**, defined as the mass-free wave in mass-free space, need not be zero. It is a force-free precursor field wave set in space with no charges present. Further, for a given $q \neq 0$, if $\mathbf{E} = 0$ and thus precursor field interactions are absent, then $\mathbf{F} = 0$, but $q \neq 0$. For that reason, as Jackson admits, the *observable* **E**-force (i.e., in charged matter q) has to involve the product of charge and field. But the notion — almost universally *assumed* by universities teaching CEM/EE — that a force field somehow exists in mass-free space, Jackson admits, is totally false.

The energy flux of the vacuum is enormously dense and comprises the scalar potential of Einstein's spacetime itself. This energy decomposes into Whittaker harmonic sets of bidirectional EM longitudinal phase conjugate wave pairs {xxi}.

It is an axiom that any mathematical model, including this one, can be engineered by a model of sufficiently higher group symmetry. The mathematics generally used to model Western classical force-field electrodynamics is U(1), a lower group symmetry model. Thus, CEM/EE can be engineered using a more sophisticated mathematical model, as the preceding discussion of precursor fields pointed out.

Our scientists, engineers, and thermodynamicists have also largely failed to grasp the implications of the gauge freedom axiom when *asymmetrical* rather than *symmetrical* regauging is used. The first requirement for EFTV systems producing more energy output than the operator alone inputs is to break Lorentz symmetry in the system's circuitry and thus to violate the Lorentz invariance imposed by the U(1) theoretical equations by which CEM is presently modeled.[1] Further, when

---

[1] See Maxim Pospelov and Michael Romalis, "Lorentz Invariance on Trial," Physics Today, 57(7), July 2004, p. 40-46 for a good discussion of present efforts to violate invariance and Lorentz symmetry.

one deals with force-free EM fields, potentials, and waves as they exist in mass-free space, *a priori* one does not have to "pay" to furnish the energy flows that are necessary to develop powerful working forces in matter. Instead, one only has to deliberately direct *nature's own energy flows*, exhibited by the internal Whittaker decomposition of any static EM potential, field, or wave in space; the energy is already pouring freely from every charge.

Force is an ongoing effect of a more primary ongoing precursor interaction between dynamic spacetime (and dynamic virtual particle flux or VPF of the vacuum) with mass. Force is not a fundamental cause. Every EM field or potential in space is in fact already a great set of ongoing free EM energy flows, steadily flowing from the associated source charge(s). Any "energy" being lost to provide that set of energy flows is furnished by the reordering of disordered virtual energy absorbed by the source charge from the seething vacuum. Since the totally-disordered vacuum is already at maximum entropy state, at equilibrium, any change in that state is a negative entropy change a priori. Using such thus involves negative entropy engineering. (The terms "precursor engineering" and "negative entropy engineering" are essentially synonyms.)

Thus precursor engineering advances beyond the present notion of four primary forces of nature. Instead of four primary forces, precursor engineering deals with four "brands" of precursor wave sets and energy flows, each kind interacting with a particular kind of mass to produce one particular kind of "fundamental force". *The Whittaker decomposition for an EM force field in matter actually gives the energy flow set, which is the form of the actual precursor EM energy field in space.* Envelope summation never occurs until the precursor energy interacts with charged matter.

In theory, by asymmetrically directing these Whittaker precursor energy flows, one can generate forces of any strength and pattern in charged matter. One can even shape the internal dynamics (to be explained later) of such forces. We refer to forces created in this manner as "engines". The energy for creating these engines is present in nature; one only needs to furnish switching and control. By paying a small energy "cost," one can direct and structure the desired patterning and dynamics of large precursor flows of energy in space. The interaction of these "precursor engines" with charged matter can produce very powerful forces in charged matter.

In hard military terms, precursor engineering (that is, Soviet energetics) provides an enormous and ubiquitous force amplifier that may be put to startling use on the battlefield.

Very large energy effects in physical matter can thus be obtained for only a small investment in switching and control of nature's ubiquitous free energy flows. Conservation of energy is not violated, since the charges furnish the excess energy utilized, in their continuing extraction of real EM energy flow from the vacuum's fluctuations.[2] Thermodynamics is not violated, because these precursor energy flows are nonequilibrium steady state systems, with continuous free energy input from the vacuum's subquantal fluctuations and with continuous free energy output in the form of the precursor energy flows.

This is *negative entropy engineering,* the ultimate "far from equilibrium" approach that totally violates the present flawed Second Law of thermodynamics {xxiv}.

Observable forces and force fields do not exist in empty space. They exist only in matter, because matter is a necessary component of force. Subquantal energies fluctuating in the vacuum of space are not observable until they produce state changes in matter. Our fundamental mechanics greatly errs in assuming a separate vector force in mass-free space, acting upon a separate mass. No such "separate force in mass-free space" exists, and none can exist, since mass is a component of force by $\mathbf{F} \equiv d/dt(m\mathbf{v})$. If $m = 0$, then $\mathbf{F} = \mathbf{0}$. When there is no mass, there is no force. As Nobelist Feynman stated: "Everything we know is only some kind of approximation, therefore, things must be learned only to be unlearned again or, more likely, corrected."

Speaking specifically about force, Feynman stated {xxv}: "...in dealing with force the tacit assumption is always made that the force is equal to zero unless some physical body is present... One of the most important characteristics of force is that it has a material origin..." "If you insist on a precise definition of force, you will never get it!"

But then, facing Feynman's dilemma of defining a force, what kind of electromagnetics exists in force-free space, prior to the presence of mass and thus prior to the formation of forces and force fields assumed by the present CEM/EE? Speaking of the electric field from a source positive charge in space, and the responses of charged masses placed in that field and interacting with it, Feynman stated {xxvi}:

> "...the existence of the positive charge, in some sense, distorts, or creates a "condition" in space, so that when we put the negative charge in, it feels a force. This potentiality for producing a force is called an electric field."

Feynman also pointed out that we really did not know what "energy" was. In the particle physics view of energetics, what basically exists in mass-free spacetime is the disintegrated and disordered virtual particle flux (VPF) of the vacuum. *Energy* in said space means an ordered change in that VPF. In the general relativity view of energetics, the basic space is "spacetime" which now must be considered highly structured and highly dynamic. *Energy* in said spacetime is a change or distortion in its curvature.

---

[2] In this book we give the exact mechanism by which the charge does this.

All of the so-called "energy in space" — such as a field or a potential — may be thought of as an organized change of the *virtual state energy* and curvature of spacetime. The "enormous" vacuum energy (using the term "enormous" as if the energy were already coherently integrated) is actually *disintegrated into very tiny virtual bits of energy and ordered patterns in that energy*. The energetic vacuum consists of intense and dense subquantal fluctuations! An EM field, potential, or wave in vacuum (i.e. in empty space) is actually an organization (ordering) and dynamics imposed upon and in the basic disordered VPF and virtual energy. The fundamental field or potential is thus an "organizing pattern" of ongoing energy flows, and must already include negative entropy, since it reorders the disordered virtual energy of the vacuum *a priori* {xxvii}.

Any persisting pattern (organization) of EM change in a region of the VPF — once it interacts with observable charged matter — can be and is integrated to the quantum level so that it becomes observable EM change and produces force and force fields in the charged matter with which it interacts.

There is no observable force *a priori* until a pattern of virtual change in a region of the vacuum (a *vacuum engine* or *precursor engine*) is reacting with observable matter and being integrated to observable quantum level in that interaction. We have explained that integration process elsewhere {xxvii, xxviii} as demonstrated by every charge in the universe.

If charges did not reorder and integrate disordered virtual state energy changes to observable state energy changes, there could not be any observable EM forces in nature, or any observable masses in nature.

How the Notion of Transverse EM Waves in Space Originated

To explain sets of longitudinal EM waves in the vacuum, and also to explain their detection as transverse waves in a receiving conductor {xxix}, consider the longitudinal flow of Drude electrons in a conductor, which is known to be severely restrained. While the electrons in the Drude gas may individually move at greater velocity, their net flow longitudinally down the conductor is the *drift velocity* {xxx}, usually on the order of a few centimeters per second in a typical bench circuit with small voltages and currents {xxxi}, very much less than the speed of the field or potential down the wire. Hence there exists an effective restraint of the electron's spinning gyro axes in the longitudinal or current-flow direction along the conductor.

To a first order, when the precursor longitudinal EM force impinges upon them, the Drude electrons — due to their continuous spin — act as gyros with longitudinally restrained axes. They occasionally slip a bit *longitudinally*, but are immensely freer to move *transversely*. So they easily precess with lateral motions that can be very large and/or very fast indeed.

The intense lateral precession of the "Drude electron as a gyro" quite well demonstrates the overall longitudinal nature of the *causative* disturbing precursor agent(s), with the longitudinal impulse force strongly resisted by the longitudinal restraint force, and with most of the electron's *effect force* being the resulting gyro-electron precession force at right angles.

Even when electrons are in empty space and not in a conductor, their mass inertia and spin results in a similar much larger precession at right angles to the longitudinal disturbing forces of the precursor force-free fields and waves.

The observation of the lateral electron precession waves of these precessing Drude electrons in our instruments is the accepted measurement of the so-called transverse EM waves in vacuum. The measurement does not measure the oscillation direction of the causative precursor EM wave in vacuum at all, but only its oscillation effect created in the precessing Drude electrons of the charged matter. There are really no transverse EM waves in mass-free space, albeit we seem to be stuck with such a model since the founders of classical electrodynamics assumed the material ether filling all space, which they modeled as matter waves (force field waves) in space.

If there **were** such a material ether, then there would be matter and charge at every point in space, and there would indeed be transverse charge precession EM waves in this "ethereal but real matter" filling all space. The CEM/EE model of electrodynamics and its equations have implicitly propagated the erroneous assumption of a material ether and the misinterpretation of what our instruments are measuring, from at least 1865 to the present day. Yet the material ether was falsified over a century ago, in 1887. The false assumption of the material ether and the resulting false assumption of the transverse EM force field wave in space continues even after Drude's fundamental work — just as the erroneous notion of a separate force vector in space acting separately upon a mass is still propagated in basic mechanics.

It is astounding that the NAS, NSF, NAE, Doe, national labs, and universities blithely continue to propagate such known errors in major models, even after eminent scientists such as Wheeler, Feynman, and many others have pointed out that there are no forces or force fields in space.

We know of no other Western analyst who has pointed out the lateral precession of Drude electrons resulting from their longitudinal restraint to a slow drift velocity. This phenomenon appears to prove that the EM wave sets in matter-free space are longitudinal rather than transverse, precisely as Tesla stated.

We pointed out this anomaly in the erroneous "force in empty space" theory for years, to little avail. The Soviets not only discovered it but also highly weaponized it shortly after WW II, in their secret superweapons science called "energetics".

CEM and EE Still Erroneously Assume the Old Material Ether

The misinterpretation of the detected transverse EM force field waves as actual force-field waves in a material space has been perpetuated erroneously since the inception of classical electrodynamics, and it continues today. The entire U.S.

scientific apparatus — including the National Academy of Science, National Science Foundation, National Academy of Engineering, Department of Energy, national laboratories, and universities — continues to bury its head firmly in the sand, covering itself with the old luminiferous material ether.

Because of its importance, we reiterate that those *measured* waves in our measuring instruments *actually* are integrated transverse EM force field *effect* waves in and of the interacting Drude electron material medium in the conductors of the intercepting instrument. But the *causative* interacting EM field entities in the vacuum are themselves force-free *longitudinal* EM wave set disturbances of the curvature of spacetime (general relativity view) and of the local VPF of the vacuum (particle physics view). *Else the theory of gyro precession is voided by every Drude electron and transverse EM force field wave detection in Drude electron gases.*

The EM force fields so blithely assumed by our present electrodynamics absolutely do not and cannot exist in mass-free space, since force only exists in a mass system, and then only during its interaction with force-free precursor fields. Because of charged particle spin, the elemental responses of the charged mass system are precessions at right angles to the disturbing force-free fields. Neither do transverse EM "effect" waves exist in mass-free space, but only in charged mass systems (such as the Drude electron gas) engaged in an ongoing interaction with the longitudinal precursor waves. Here also, the elemental response of the mass system is at right angles to the force-free fields of the disturbing precursor wave set.

The primary, longitudinal force-free fields in massless space are the more fundamental EM fields, as Tesla had realized. Tesla actually began the process later partially captured by Whittaker {xxi, xxii}, which was taken to fruition after WW II by Soviet weapon scientists as the new weapons science of *energetics*. It is what the present author is calling *precursor waves*, *precursor fields*, *and precursor engineering*, in trying to penetrate the mind-fog barrier generated by the "transverse EM forcefield wave in space" idea so branded into scientific and engineering minds.

Our Archaic and Flawed Situation in CEM/EE Today

Sadly, electrical engineering departments still erroneously teach EM force fields and EM transverse EM force field waves in space — a notion which still assumes the old luminiferous material ether {xxxii} that was falsified in 1887 by the Michelson-Morley experiments {xxxiii}.

Reiterating: Jackson, one of the superb electrodynamicists of our day, at least admits that the problem exists and that electrodynamicists continue to ignore it. He states {xxxiv}:

> "Most classical electrodynamicists continue to adhere to the notion that the EM force field exists as such in the vacuum, but do admit that

> physically measurable quantities such as force somehow involve the product of charge and field."

In summary, one does not have electromagnetic force until the "force-free precursor field in space" is interacting with charged mass. It can easily be seen by examining the simple standard EM equation $\mathbf{F} = \mathbf{E}q$, and writing it as $\mathbf{F} \equiv \mathbf{E}q$ so that it becomes a true definition of the electromagnetic force field. Now let $q = 0$, as it would be in space. In that case, $\mathbf{F} = 0$, even though $\mathbf{E} \neq 0$. In that identity, the actual **E**-field in space is force free, and force is the ongoing interaction — of the force-free **E**-field in space — with charged mass q. Given q and its interaction with force-free **E**, that ongoing interaction is the rigorous definition of — and generatrix for — electric field force itself. What is missing from the identity and therefore the definition is that **F** and **E** are at right angles to each other because of the spin of the charged particles comprising q.

Again, let $q = 0$, while **E** remains nonzero. In that case, there is no ongoing interaction of **E** with charged matter, since *there is no charged matter*. Therefore the *effect* of such a reaction — force **F** — goes to zero because *there is no interaction* and therefore *there is no effect of an interaction*. So the equation should really be written as an identity, to show clearly that electric field force is produced by, and exists as, *an ongoing interaction of the force-free **E**-field in space with charged matter q*.

The very concept of "electromagnetic field" must therefore be separated into two parts. Using "the **E**-field" as an example, $\mathbf{E}_M$ refers to the **E**-field interacting with mass, and $\mathbf{E}_S$ to the **E**-field in empty space. Using this notation we might state that $\mathbf{E}_M = f(\mathbf{E}_S \times q)$. In short, if there is any chance for misunderstanding, the writer/lecturer should clearly explain whether he is referring to the force-free **E**-field $\mathbf{E}_S$ in space, or to the force **E**-field $\mathbf{E}_M$ in matter. Indeed, the CEM/EE model itself should be altered to clearly show that difference.

The sharp-eyed reader will note that we may have (hopefully successfully!) indicated the solution to Feynman's force definition problem, at least electromagnetically: *An electromagnetic force in static charged matter is identically the ongoing effect (in the static charged matter) of an ongoing interaction of the force-free EM field in space with that static charged matter.*

Eminent scientists such as Feynman, Wheeler, Lindsay, Margenau, Bunge, and many others have pointed out that there are no force fields in space. But our electrical engineering departments and classical Maxwell-Heaviside electrodynamicists refuse to change and correct the horribly flawed CEM/EE model. Repeating for emphasis, Feynman {xxvi} stated it this way:

> "...the existence of the positive charge, in some sense, distorts, or creates a "condition" in space, so that when we put the negative charge in, it feels a force. This potentiality for producing a force is called an electric field."

In other words, a true $\mathbf{E}_S$-field is a "force-free condition in pure mass-free space;" it has only the *capability* to interact with a charged mass and thus create force in that ongoing interaction. Feynman understood that "true fields" were naught but conditions in space itself, and changes in space itself. He clearly understood that only the interaction of the field-in-space with charged mass — *should* some be introduced — produces EM force fields in that charged mass.

So our present electrical engineers have absolutely no notion at all of what a force-free EM field in mass-free space really is. *They have never even calculated a real EM-field in space, and they do not do so today — nor do their professors and their textbooks, though all profess to do so.*

One is reminded of Tesla's acrid remark[3] when he stated:

> "The Hertz wave theory of wireless transmission may be kept up for a while, but I do not hesitate to say that in a short time it will be recognized as one of the most remarkable and inexplicable aberrations of the scientific mind which has ever been recorded in history."

Tesla also was well aware that energy could be had from space, for the taking, and he was at some pains to express it. In 1900 he stated:

> "Whatever our resources of primary energy may be in the future, we must, to be rational, obtain it without consumption of any material."

Nearly a decade earlier, he had formally[4] stated:

> "Ere many generations pass, our machinery will be driven by a power obtainable at any point in the universe. This idea is not novel... We find it in the delightful myth of Antheus, who derives power from the earth; we find it among the subtle speculations of one of your splendid mathematicians...Throughout space there is energy. Is this energy static or kinetic? If static our hopes are in vain; if kinetic — and this we know it is, for certain — then it is a mere question of time when men will succeed in attaching their machinery to the very wheelwork of nature."

Tesla himself was stopped from doing that very thing, by powerful vested interests who realized that for the financial cartels involved in energy, energy must continue to come from fuels and materials, etc. Today, similar cartels of very great power still control which way science thinks and researches, and still effectively destroy any scientist who dares to seek energy from the vacuum, and Tesla's nonmaterial energy source "obtainable at any point in the universe".

---

[3] Nikola Tesla, "The True Wireless," Electrical Experimenter, May 1919.

[4] Nikola Tesla, in a speech in New York to the American Institute of Electrical Engineers, 1891.

The primary factor responsible for successfully suppressing most attempts to scientifically research EFTV is the scientific community's continuing insistence on the absurd notion of force fields in space.

Other Severe Foundations Errors in Our CEM/EE

Indeed, besides the material ether (falsified over a century ago), the inane CEM/EE model still erroneously assumes a flat spacetime (falsified since at least 1916) and an inert vacuum (falsified since at least 1930). It also assumes that every EM field, potential, and joule of energy in the universe was freely created from nothing at all by the associated source charge, which the theory begrudgingly admits continuously emits real observable EM energy without any energy input. In so doing, the CEM/EE model implicitly assumes the universal violation of the conservation of energy law.

Every EE department, professor, and textbook in the United States unwittingly advocates the total violation of energy conservation, without even recognizing it. So do our National Academy of Sciences, National Academy of Engineering, National Science Foundation, Department of Energy, great national laboratories, universities, our intelligence analysts — and the entire Taliban-like professional Skeptic community — by not pointing out the known terrible errors in classical electrodynamics and electrical engineering, and by not correcting the flawed model!

The concerned reader should make a determined effort to find any single textbook that presents and lists the foundations assumptions in the CEM/EE model, and discusses which are still true and which have been falsified by the progress of modern physics. The reader will not find any such textbook, nor will he find any project to correct the textbooks, in any of the previously named scientific groups and organizations.

The Development of Scalar Interferometry

There are two papers by E. T. Whittaker, one in 1903 {xxi} and the other in 1904 {xxii}, that detail the beginning of scalar interferometry. Shortly after WW II, these two papers were a starting point in the Soviet Union scientists' development of a secret weapons science called *energetics,* which led to the development of powerful *scalar interferometers* that are now possessed secretly by at least 10 nations worldwide, and even by the Japanese Yakuza. For a deeper understanding of scalar interferometry, a higher group symmetry electrodynamics must be used {xiv}, instead of the Standard Model's limited U(1) electrodynamics.

After leasing large strategic scalar interferometers in Russia, the Yakuza took the technology back to Japan and set up production facilities of their own. This was *not* an act of the Japanese government but of the Japanese Mafia in concert with a remnant of the Aum Shinrikyo (under its new name, Aleph). The Yakuza is very powerful, and it penetrates the Japanese government and also every large Japanese company. *All dealings with any large Japanese company inherently involve the Yakuza, due to their penetration into all aspects of Japanese society and business.*

In 1997, then Secretary of Defense William Cohen — at a conference on Weapons of Mass Destruction (WMD), held in Georgia under the auspices of Senator Sam Nunn — publicly confirmed this kind of electromagnetic weaponry without the technical details. He also confirmed the weather engineering, climate control, induction of earthquakes, and stimulation of volcanoes into eruption. Secretary Cohen stated {xxxv}:

> "Others [terrorists] are engaging even in an eco-type of terrorism whereby they can alter the climate, set off earthquakes, volcanoes remotely through the use of electromagnetic wave. ... So there are plenty of ingenious minds out there that are at work finding ways in which they can wreak terror upon other nations. It's real, and that's the reason why we have to intensify our efforts, and that's why this is so important."

This was the first time (and to my knowledge, the only time) a high U.S. government official has openly confirmed the superweapons. So obviously certain parts of our government do know about such weapons, but there has been little or no communication of that to the public, save by Secretary Cohen's illuminating statement.

The news media at the time — panting heavily after the juicy Clinton-Lewinsky sex scandal — totally ignored Cohen's epochal strategic statement. So much for the perspicacity of our news media!

The Yakuza Strategic Role in Our Present Terrorism Threat

Today the Yakuza/Aum Shinrikyo, one of our deadly foes, will be involved in the planned destruction of the United States that is scheduled to conclude about three years from now. A major facet of the plan is the destruction of the power grid in about two year's time. The incidents will start very gradually at first[5], and then escalate fairly rapidly until the power system is in ruins. There are already sufficient terrorist assets inserted into the U.S. — being held ready and waiting — to destroy the nation, particularly when one considers the damage potential of scalar interferometers. A single large interferometer can, for example, disable the entire power system and keep it nonfunctional by causing power surges at will. It can destroy power plants, grids, distribution stations, refineries, etc. Or it can keep the system failing and nonfunctional by damaging or destroying electronic controls.

---

[5] The 2004 hurricane passing through the Gulf of Mexico did substantial damage to the oil and gas pipelines and some of the oil rigs. That decreased our daily supply of oil and gas, was been partly responsible for oil prices going above $50 per barrel, and still plays a role in their continuing to rise.

One keeps hoping that the Administration — whether Republican or Democratic — will alert the American public to this possibility, but nothing has been said by officials of either party. *Too many high level government people take the attitude that the American public just could not withstand such information.* Further, within the government the *entourage* {xxxvi} seems to have again largely suppressed the information that Defense Secretary Cohen knew and publicly confirmed.

The Yakuza strategic threat to the U.S. is one of the most potent we face today, and it increases year by year. That threat — as is much of the organized international terrorist threat — is being generally coordinated from the old die-hard Communist faction of the FSB/KGB. The Yakuza strategic threat is also being generally ignored by our intelligence analysts, yet the Yakuza are protégés of the same dominant FSB/KGB faction, the one which possesses, controls, and utilizes the Russian energetics superweapons, never having allowed them to pass into the regular Russian armed forces. Thereby the communist faction retains power and will do so.

President Vladimir Putin, from the younger faction of the KGB, would like to reach an accommodation with the U.S. However, the Russian communist faction, which remains dominant, prohibits him from doing so. Accordingly, Putin has been forced to reinstitute harsher state control measures and take control of domestic oilfields from private industry. Instead of becoming the "trusted cheap oil supplier" to the United States as Putin wished, Russia has allowed China to invest in developing Russian oil fields. This appears to be part of the new alliance among Russia, China, and India.

Suppression of "Energy-from-the-Vacuum" Electrical Systems

Electrical power systems freely taking their energy from the vacuum rather than from fuel are suppressed worldwide by several large and powerful cartels — including in the U.S. scientific community — and for some time the Yakuza has suppressed all EFTV systems successfully developed in Japan.

As part of their scheduled plan to catastrophically collapse the U.S. economy by obliterating our power supply, the Yakuza have suppressed several of these legitimate Japanese overunity electrical power systems — such as the Takahashi magnetic Wankel engine {xxxvii} and the Kawai motor {xxxviii} — to prevent them from being marketed.

Were these already-developed Japanese systems allowed on the market, they would quickly resolve the world energy crisis, and enable the dismantling and rapid replacement of the horridly vulnerable, centralized U.S. electrical power system. Needless to say, the availability of such systems would cleanse the biospheric contamination from energy-related operations and developments, alleviate the march toward global warming, etc.

All the field energy and potential energy used in every EM circuit already comes directly from the vacuum, and not from cranking a generator shaft or dissipating chemical energy in a battery. The energy is extracted from the vacuum itself by the system's source charges and dipolarities. And yet our own scientific community perpetuates a model assuming that the field energy from the charges — and therefore the available usable energy in the circuit — is freely created out of nothing at all, while also erroneously teaching that it is the mechanical power turning the generator shaft which powers the electric distribution line and distant loads.

We do not have to rediscover how to extract usable EM energy from the seething vacuum! Every charge in the universe already continuously does that, and it continuously gushes forth real EM energy extracted from its ongoing "ratcheting" vacuum interaction. All we have to do is learn how to properly intercept, extract, and use the already-real energy flows from source charges and dipoles. This has to be done by uncovering and eliminating the limitations unwittingly incorporated into the CEM/EE model — such as the ubiquitous closed current loop circuit containing the source dipolarity (the source of potential wired into the external circuit as a load to be continually destroyed — and such as insisting on invariance and Lorentz symmetry of the model because of the "mathematical beauty" in symmetrized equations and an invariant system. "Invariance" refers to the notion that no new net observable is permitted when the equations are regauged. To do that requires symmetrical regauging, which kills any modeling of asymmetrical receipt of excess energy from the active environment. The Lorentz symmetry of our inane electrical power circuits thus guarantees that the much simpler Lorentz-invariant equations can be used to describe the deliberately crippled system's operation. It also guarantees that the system will not electrically perform with a coefficient of performance (COP) > 1.0.

The scientific and industrial and financial entourages love it, and will do whatever is required — destroy careers, withhold funds and tenure, or even kill the researchers — to keep that status quo. Hence it is not accidental that, in a world filled with an uncountable number of free energy generators called "source charges", our vaunted scientific community still opposes energy from the vacuum systems and will not fund or allow research on them. Their position is the ultimate irony, because all of the energy driving every EM system ever built has been extracted from the seething subquantal fluctuations of the vacuum!

On the other hand, by deliberately avoiding the Lorentz-invariant equations, one can violate the Lorentz symmetry, resulting in *asymmetrical* circuits and systems that collect excess energy (i.e., asymmetrically regauge themselves) from the active vacuum's interaction with all the circuit's charges. *Asymmetrical* regauging freely produces a *net nonzero force field.* A properly designed circuit can utilize this force field to dissipate the previously-collected "free potential energy" in external loads to power them almost "for free" (as far as any cost to the operator for the energy dissipated). No law of physics is violated, and no valid law of thermodynamics is violated (once the erroneous Second Law is corrected, and is no longer the oxymoron and "half law" it presently is).

Sometimes simple equations contain profound truths. Because a potential (envelope) is actually a set of ongoing EM energy flows, then from any fixed source of potential V, any amount of energy W can be freely collected on intercepting charges q, *given sufficient q and also dq/dt = 0*. The simple equation is W = Vq. Further, by only increasing V, while pinning a fixed amount of intercepting charges q so that i = dq/dt = 0, *any amount of potential energy density V can be freely produced in the system by work-free voltage amplification alone.* Regauging is changing the potential(s), nothing else.

From such simple equations, one can see that a circuit can self-regauge and take on extra potential energy freely, from any external source of potential energy providing the necessary potential energy flow, without having to do work — *so long as the source of potential is not deliberately destroyed.* This is done in a two-phase process. In the first phase, dq/dt is made to equal 0 by pinning the charges q to prevent current flow. In this phase the pinned charges are freely potentialized. One can potentialize as many charges as desired without draining the external source of potential, because the source's dipolarity is not reduced. In the second phase, the source of potential is disconnected, the circuit is re-closed and the charges q are unpinned. Now the circuit freely dissipates the previously-collected energy to power loads "for free". The circuit can perform this two-phase process repetitively and continuously because dq/dt = 0 during the potentialization process.

In short, the external source of "static" potential will last indefinitely, so long as current is not forcibly pushed up through its dipolarity's back emf. *Via a flashlight battery or electret as the "external potential source," under proper circumstances sufficient energy could be extracted from the vacuum to power New York City, at least in theory.*

We do not have to learn how to extract energy from the vacuum. We only have to learn how to more properly use the energy freely flowing from charges already doing so, and build and use circuits that do not keep killing the source of potential energy flow faster than they power their external loads. The energy we input to the shaft of the generator is to continually restore the destroyed dipolarity, that we intentionally build our "external systems" to do, faster than they power the external loads. We pay the power company to engage in a giant wrestling match inside its generators and lose.

Our classical electrodynamicists and electrical engineers have missed the fact that, since every EM field and potential is a pattern in the continuous flow of EM energy from the associated source charges, *the field or potential is itself a set of continuous free EM energy flows.* To see it and prove it, simply decompose the field or potential by Whittaker's methods {xxi, xxii}.

Members of our company, CTEC, personally witnessed the Yakuza suppression of Japanese EFTV electrical systems during our negotiations for rights to market Kawai's self-powering Kawai motor right here in Huntsville, Alabama in 1996. The board of directors and the author, as chief executive officer, had reached agreement with Kawai late one Thursday afternoon. The Yakuza arrived by private jet from Los Angeles that night. The next morning, a stunned and totally demolished Kawai no longer controlled his invention, his company, his life, or further developments. *If not for the Yakuza's intervention, CTEC would have placed self-powering Kawai motors on the world market by mid-1997.*

Such self-powering electrical systems are no more mysterious than windmill-powered generators or solar cell array power systems. In each case, the active vacuum environment furnishes all of the energy required. Even though any system has losses and its overall thermodynamic efficiency cannot exceed 100%, the coefficient of performance can equal infinity when the external active environment, not the operator, is the source of all of the energy collected.

Energy-from-the-vacuum systems to replace our centralized electrical power grid would take nearly-free energy directly from the local vacuum, needing no external fuel, pipelines, fleets of tankers, railroad coal cars, refineries, etc. Most of the massive and expensive *support structure* for our bloated centralized electrical power system would gradually be dismantled as it was replaced by a decentralized EFTV system. Our present dependence on foreign oil and gas would vanish forever, accompanied by a dramatic reduction in environmental pollution and a dramatic uplift to the economy.

The U.S. electrical power system would no longer be remarkably vulnerable. Replacement of most of the centralized power system would be of enormous economic benefit and of enormous strategic importance. Presently, we have some two years to get a substantial percentage of that task done, in precisely that manner. Otherwise, we will see the centralized electrical power system destroyed, followed shortly thereafter by the catastrophic collapse of the U.S. economy. And then will immediately follow the methodical destruction of the United States — every last man, woman, and child. At least those are the plans of the enemy waging the asymmetrical war against us, and soon to spring up and proceed internally inside the United States.

The resulting benefits of such a centralized, fuel-free system to the taxpayer and to the long suffering destitute peoples of the world would be incalculable — but the incredible loss of revenue to the large energy cartels now in power worldwide *would be directly calculable!* Such powerful organizations do not surrender their vast empires and dominant positions without a major struggle.

So certain elements of those cartels already suppress "free energy" developments worldwide, and they have been doing so since Stubblefield {xxxix} (even before Tesla). Those who engage in energy from the vacuum research, if successful, can easily "meet with a sudden suicide" on the way to the supermarket, unless they exercise exceeding care. Or one can be hit in the midst of a crowd by an ice dart dipped in curare, so that one convulses and dies right there with a "seizure" and massive heart and brain damage.

Or heroin can be planted on the premises secretly by an intruder. Then the regular narcotics police are "tipped off". They raid the place, the dogs sniff out the large cache of heroin, and one has "been caught red-handed" as a dirty old dope dealer. One winds up behind bars for 20 years or so, wondering how it all happened.

Or one encounters a "throw-away" assassin, who has been brutally conditioned. After he successfully fixates on his target, then later he will march right up to the targeted individual in a public restaurant, pull out a pistol and empty it at point blank range into the target, killing him. He will remain right there until the police come and haul him away. And even electric shocks on his testicles will not elicit any useful information from him.

Or, more simply, one just meets with a fatal auto accident.

Or one gets hit with a microwave "shooter" whose wavefront has been carefully modified by the Venus ECM (electronic countermeasures) technique, to dramatically disrupt the receiving heart by throwing it into violent and uncontrolled fibrillation. The target falls, goes into convulsions, thrashes a bit, and expires with a legitimate massive stroke, heart attack, or both. The smaller short range shooter is about the size of a small paperback textbook, and fits inside the assassin's coat pocket. A larger longer range shooter is about the size of a bazooka. It will shoot right through walls and windows, killing a person inside the building (after locating the target's body heat through the wall or window with an infrared device).

Or a professional hit man kills the person, making it appear that some hoodlum committing a robbery or some such did the crime. Recently Dr. Gene Mallove, an excellent alternative energy and cold fusion researcher and a highly dedicated scientist, probably was murdered by this latter type of professional hit. It was probably arranged by the entourage of the powerful consortium (both industry and science) planning to rip off the U.S. public for more than a trillion dollars, to build hundreds of pebble bed nuclear reactor power plants. To do that, they need helium, which is in short supply and expensive. But in "cold fusion" there is a nuclear reaction which generates helium {xl}, and it can be developed to do it copiously and cheaply. So the cartel needs to appropriate that reaction from cold fusion, and pretend it is hot fusion, while still keeping cold fusion suppressed. Mallove, the great champion of cold fusion, would have prevented that, and he had the respect and connections to do it. Obviously, Mallove had to go. No one would have believed suicide, so it had to be more mundane. Hence a "lowly burglar" beat him to death.

Those are just a sampling of the many methods — professional clandestine assassination, jailing, suppression — currently used against some EFTV researchers.

Unfortunately, President Bush — with a serious energy crisis approaching and with continuing bum scientific advice from our scientific community — has been persuaded to back that big nuclear power program. So the cartels in industry and science that are grabbing all the research funds for big nuclear power, hot fusion, hydrogen economy, etc. are in fact preventing any appreciable research in the one area that can do the job: energy from the vacuum. Planning and funding the building of big nuclear power plants that will go on line in about 5 to 7 years, to feed a distribution system that will be destroyed two years hence, in a United States that will be destroyed within a year following, simply will not do the job. Instead, it will exacerbate the problem. *It will, however, guarantee that when Rome burns it will burn completely*. Presently that is the main function of our power industry and our scientific community, who still have no notion of the terrible threat and destruction bearing down on this entire nation.

But to seek out what our own science is doing in EFTV research, simply visit the websites of the National Academy of Science, National Science Foundation, Department of Energy, national labs and universities, and search for any real EFTV research program or solution that would enable our survival in this three year period we have entered. Cut through the rhetoric and hype for any real content that would solve the problem. There is none, and there is not going to be any.

The sites and organizations are still blindly accepting the very seriously flawed electrical power engineering, the material ether, force fields in empty space, flat spacetime, inert vacuum, and other gross scientific errors. The reader will find that "business as usual" and the interests of the powerful scientific cartels — hot fusion, nuclear power plants, fuel cells, etc. — are being pursued, with a little sop to the environmentalists in such things as solar cell array power systems, windmill-powered generating systems, etc.

To sum it up: Not only is our scientific community fiddling while Rome prepares to burn, but it is also ensuring that Rome burns completely.

Yakuza Production of Portable Scalar Interferometers

The Yakuza now produce their own scalar interferometer weapons in their own facilities in Japan, including small, portable versions. This Yakuza activity has no relationship to the Japanese government. Small versions of these Yakuza weapons which can fit into a small sports utility vehicle (SUV), and undoubtedly some are also already inserted in the U.S., to be used a few times during the next two years against our refineries, nuclear power plants and hydrocarbon-fueled power plants, oil fields and Gulf oil rigs, control systems and substations for the centralized power grid, pipe line hubs, long high voltage distribution lines, etc. In short, *the Yakuza can already lay down the U.S. electrical power system, for months at a time or permanently — and it can do it easily, quickly, and at will.*

Even normal cyberwar techniques will do it for extended periods, as is well known by cyberwar specialists. For example, quoting from a PBS interview with Joe Weiss {xli}:

> Question: "So just put it all in perspective. What's the worst-case power scenario, power we're talking here — power lines, power grid?"
>
> Answer: "Absolute worst? I won't even say absolute, but a very worst case could be loss of power for six months or more."
>
> Question: "Over how big an area?"
>
> Answer: "Big as you want."
>
> Question: "Is that a possibility?"
>
> Answer: "Yes."

Presently we are in the two year "slowly escalating" initial operations phase of the centralized asymmetric warfare plan for the destruction of the United States.

Then, beginning about two years from now, the final operations phase will swiftly occur, and the already damaged entire U.S. centralized electrical power system will be destroyed quickly and rather permanently. The destruction will be by combined attacks, including cyberwar, with a guaranteed destruction furnished by Yakuza tactical and strategic interferometers, portable EMP shooters, etc.

The destruction of the United States, with the aid of the Yakuza, is coordinated by centralized planning under the old die-hard Communist faction of the FSB/KGB, which controls the Russian Federal Security Service and the Russian superweapons. Possession and operation of the Russian superweapons ensures this faction's dominance.

Russian President Vladimir Putin comes from the younger, more modern faction of the FSB/KGB, which would very much like to reach an accommodation and become the United States' trusted cheap oil supplier. This arrangement would benefit every level of the Russian economy as result of pumping in U.S. development and royalty money.

However, supplying copious Russian oil to the United States contradicts the Communist faction's fervent plans to totally destroy the United States. Thus Putin has been stymied to a large extent {xlii}. Instead, Russian oil shipments to China, over land by rail, were stopped, stimulating China into a more aggressive stance in other world oil markets such as the Middle East. Consequently, China bought in to the Russian oil fields, after the largest oil consortium in Russia came under legal attack and direct interference by the State, with the State eventually taking over the oil fields under the guise of a State protégé company. Indeed, a new alliance between China, Russia, and India bodes ill for our access to energy and for other things as well.

The old Cold War has not gone away, contrary to popular conception. Instead, it has just been turned into a more sophisticated kind of war, called "asymmetric" warfare. The same "old dogs" are still calling the shots, like a tired old black and white movie from the 1930s. Hence our asymmetric war problems (i.e., our *War on Terrorism* and the war of terrorism on us) will continue. As an example, Putin, caving in to pressure, again centralized and tightened the control of the Russian government, setting aside the reforms of open election that had been partially implemented {xliii}. In addition, he had to seize the oil fields and assign them to a puppet company controlled by the Russian state.

Afghanistan, Iraq, Al Qaida, and Intended Muslim Jihad

The international terrorists are also being subtly and not-so-subtly manipulated, and we will see the anti-American polarization in the MidEast continue and intensify. Many foreign Arab fighters are fighting in Iraq and Afghanistan right now. Their number and organization — and centralized coordination — will continue to increase.

There are still serious nuclear proliferation problems to be faced in North Korea, Iran, and elsewhere. Iran has defiantly announced it is resuming uranium enrichment and nuclear power plant development. Short of a few ineffective sanctions or attacking Iran, it seems that little can be done {xliv}. However, clear signals from Israel emerge that eventually such will not be tolerated, and Israel will strike if necessary.[6] North Korea remains defiant and almost certainly already has a small number of nuclear weapons. Pakistan and India — both nuclear powers — continue a stand-off and are often eyeball to eyeball. With the alliance of India to China and Russia, that "eyeball to eyeball" confrontation could be aggravated at will.

Israel is extremely concerned about the path Iran is taking to achieve a nuclear weapons capability. The Israelis, having purchased several hundred "bunker-busting" bombs from the U.S., at some point may conceivably take out the Iranian nuclear facilities to prevent an unacceptable Iranian nuclear knockout threat against Israel.

The MidEast situation will further heat up, not lessen, over the next two years. It's called *bleeding the U.S. dragon* — and it is indeed economically bleeding us.

The intent of our foes is also to deliberately bring substantial factions of the Muslim world to a state of Jihad against the United States, necessitating additional commitment of U.S. forces out of country, and increasing the already-pressing

---

[6] E.g., see full page ad, by FLAME (Facts and Logic About the Middle East) in U.S. News and World Report, Mar. 28, 2005, p. 69, entitled "The Holocaust: Sixty Years Later: Has the spirit of Auschwitz been revived?" Quoting: "If Iran isn't stopped from acquiring nuclear weapons, there can be little question that, with or without the approval of the United States, Israel would have to preempt and, just as it destroyed the Osirak nuclear plan of Iraq in 1981, would destroy the Iranian nuclear installations — by whatever means it might take."

economic strain on the U.S. Quite simply we have run out of troops necessary to continue our worldwide tasks and missions. Unless a draft is reinstituted, we are also seriously challenging our reserves and National Guard. Any draft imposed is likely to reinitiate an intense polarization inside the U.S. over the war and America's foreign commitments.

Claiming that America had declared war against God and his messenger, Bin Laden has called for the murder of any American, anywhere on earth, as the *"individual duty for every Muslim who can do it in any country in which it is possible to do it."* {xlv}.

In an American national television interview from Afghanistan, Bin Laden stated his intention to kill innocent civilians as well as military, without hesitation {xlvi}:

> "It is far better for anyone to kill a single American soldier than to squander his efforts on other activities… We believe that the worst thieves in the world today and the worst terrorists are the Americans. Nothing could stop you except perhaps retaliation in kind. We do not have to differentiate between military or civilian. As far as we are concerned, they are all targets."

The <u>9-11 Commission Report</u> summarized the unrelenting attitude as follows {xlvii}:

> "Many Americans have wondered, "Why do 'they' hate us?" Some also ask, "What can we do to stop these attacks?" … Bin Laden and al Qaeda have given answers to both these questions. To the first, they say that America had attacked Islam; America is responsible for all conflicts involving Muslims. Thus Americans are blamed when Israelis fight with Palestinians, when Russians fight with Chechens, when Indians fight with Kashmiri Muslims, and when the Philippine government fights ethnic Muslims in its southern islands. America is also held responsible for the governments of Muslim countries, derided by al Qaeda as 'your agents'.… To the second question, what America could do, al Qaeda's answer was that America should abandon the Middle East, convert to Islam, and end the immorality and godlessness of its society and culture. [Al Qaeda states] '… you are the worst civilization witnessed by the history of mankind.' If the United States did not comply, it would be at war with the Islamic nation, a nation that al Qaeda's leaders said 'desires death more than you desire life'."

This perhaps explains the morose statement of Vice Present Cheney, only a month or so after 9/11, and after he had received intensive briefings on terrorism and its present conduct against the United States. Cheney stated {xlviii}:

> "The war on terrorism will not be over in our lifetime. It is different than the Gulf War was in the sense that it may never end. At least not in our lifetime. The way I think of it is, it's a new normalcy."

Bin Laden also certainly knows the importance economics plays in a war. His original wish for the 9/11 strikes was to hit several nuclear power plants etc., damaging the ability of the centralized grid to furnish power, as well as posing a possible nuclear hazard and risk. He felt he lacked sufficient control to do it, so opted for highly symbolic targets: the twin towers, the Pentagon, and the White House.

Indeed, after 9/11 Bin Laden stated {xlix}:

> "Al Qaeda spent $500,000 on the [September 11 attacks], while America, in the incident and its aftermath, lost — according to the lowest estimate — more than $500 billion. Meaning that every dollar of al Qaeda defeated a million dollars, by the permission of Allah!"

He declared his policy is *"bleeding America to the point of bankruptcy,"* and joked about the U.S. *"economic deficit... estimated to total more than a trillion dollars."* Indeed, our known total obligations are already some $47 trillion dollars, and climbing.

Welcome to asymmetric warfare, where war is war, peace is war, and there is only perpetual war, including a virulent form of economic warfare augmented by "stretching and bleeding the dragon" and by direct damage of infrastructure.

Our Analysts Missed the Yakuza's Key Role and Significant Aspects of the Centrally-Coordinated KGB/FSB Strategic Plan

Our own intelligence agencies do not appear to be aware of the key role played by the Yakuza. Nor are they aware, it seems, that the Communist faction of the KGB/FSB directs the entire asymmetrical warfare against us, so that, rather than isolated incidents, the actions of all the far-flung pieces are coordinated in a strategic plan of increasing intensity and effect. Our intelligence analysts do not appear to realize that we have entered into a 3-year Operations Phase schedule — consisting of a two-year initial "gradual intensification" period and a 1-year "maximum intensity and total destruction" period — to collapse and destroy the United States.

The strategy for the initial two-year period is to "bleed and disorganize the dragon" until it collapses in fatigue and economic chaos. Mind conversions — inserting an additional modified mind in a human, then switching control of the thought and behavior between the individual's normal mind and the implanted mind, thus altering the individual's behavior as desired — will be massively unleashed to guarantee total confusion and disorder. The plan for the third year is to violently attack and dismember the prostrate dragon, swiftly and utterly destroying the United States by totally unleashing the terrorist assets already inserted into the U.S. and waiting.

At present, "bleeding the dragon" is in full progress and being steadily increased as part of the overall strategic plan against us. We are already stretched well past the capability of our regular Armed Forces, and we are inducting many of our Reserves and National Guard forces for repeated tours overseas. We have no national draft, and a bitter societal polarization will occur if the draft is reinstated. Our national debt and expenses are escalating rapidly. Our country is hemorrhaging jobs also, further increasing unemployment and financial duress.

Hurricane and storm damages are being escalated by the increased intensity of the Yakuza weather engineering. The American dollar is propped up only by ever-increasing purchases of U.S. Treasury instruments by foreign nations such as China and Japan, which could change at any time. Several prominent Asian politicians and central bankers have recently made public statements suggesting a coordinated intent to divest dollar-denominated reserve assets and shift reserves to other currencies. An avalanche of covert dollar selling appears to have already begun.

Many destructive natural phenomena can be evoked or augmented by the scalar interferometry weapons possessed by the Yakuza. Such phenomena include volcano eruption, earthquake induction, and weather and climate engineering, as confirmed by Secretary Cohen's epochal 1997 statement. But the achievable phenomena also include tsunamis, freak waves in the open ocean, and rogue waves of numerous kind and occurrence {l}. Quoting Wolfgang Rosenthal of the GKSS Research Centre in Geesthacht, Germany on rogue waves {li}:

> "The waves exist in higher numbers than anyone expected... Two large ships sink every week on average. But the cause is never studied to the same detail as an air crash. It simply gets put down to 'bad weather'."

Even super-carrier ships are frequently destroyed by such waves {li}:

> "Over the last two decades more than 200 super-carriers — cargo ships over 200m long — have been lost at sea. Eyewitness reports suggest many were sunk by high and violent walls of water that rose up out of calm seas."

That's an average of ten per year. Indeed, a satellite test readily confirmed the occurrence of such previously unsuspected freak giant waves {li}:

> "As part of a project called MaxWave — which was set up to test the rumors — two ESA satellites surveyed the oceans. During a three week period they detected 10 giant waves, all of which were over 25m (81 ft) high."

And that was just a one-time survey by only two satellites.

Indeed, during that exact three week period while the MaxWave satellites were seeking out huge rogue waves, two large tourist liners endured terrifying ordeals from such waves but survived. The Breman and the Caledonian Star cruisers had their bridge windows smashed by 30m high waves in the South Atlantic.

It appears that at least 10 to 12 oil tankers are lost each year to giant rogue waves alone. To appreciate total shipping losses from all causes: In a single year — 1981 — 250 ships of 500 tons or greater size were lost and perished {lii}. Many of these were probably sunk by unexpected large rogue waves.

We hardly need point out that a giant oil supertanker on the ocean is also easily destroyed in a single shot by a strategic scalar interferometer. A burst of energy, placed adroitly under the surface by a distant interferometer, can also produce large waves that arise and smash a ship. This also "disguises" the incident as if it were due to anomalous large waves of natural origin.

With U.S. dependence on foreign oil and most of the oil delivery dependent on supertankers, oil continues to climb in price. In the past the U.S. government has opened the Strategic Oil Reserve a bit, to assist two hard-hit companies while they scrambled to reinstate their supply. It has also just recently squeaked through the authority to drill for petroleum in the Artic National Wildlife Refuge.

It would take very little Yakuza effort to increase the loss of oil tankers, further strangling the U.S. oil supply and dramatically increasing prices. Since the oil supply for all oil-dependent nations comes from known major oil field locations and harbor facilities, deliberate and careful reduction of the tanker fleet and facilities can place extraordinary pressure on oil's availability for several major nations at once.[7]

Coup de Grace: Strategic Strikes of Great Magnitude

Great "coup de grace" blows have also been prepared by the Yakuza, in order to *guarantee* the catastrophic collapse of the U.S. economy, followed by the swift destruction of the United States itself. Several of these involve opportunistic volcano eruptions and tsunamis {liii} that can readily be evoked.

As an example, prior to the end of 2004 the Yakuza registered one or more of its large scalar interferometers upon one of the world's largest supervolcanoes — whose calderas lie beneath or near Yellowstone National Park. The symbolic timing so close to the anniversary of the 9/11 attack should not be overlooked. The Yellowstone supervolcano erupts about every 600,000 years, and some 640,000 years have elapsed since its last eruption. Hence — geologically speaking — its next eruption is overdue, and the caldera undoubtedly harbors great pent-up stress. Presently a Yakuza finger is on the trigger of the registered scalar interferometer(s). If that finger pushes the trigger, the interferometer will gradually inject tremendous energy into the Yellowstone supervolcano's stress zone, resulting in a violent eruption. Its previous eruption, about 640,000 years ago, devastated much of North America, including most of the higher life forms.

By deliberately kindling the supervolcano into eruption, most of North America would be devastated, and particularly most of the United States {liv}. The supervolcano could eject more ash, lava, rock, and debris than the entire Grand Canyon can hold. Additional side effects would include the instant destruction of most of the agricultural food crops in North America by covering the farmland with thick ash a meter or more in thickness. Survivors would be faced with the immediate problems of starvation and fatal lung disease (Marie's disease), the degree of lung disease varying with exposure to silicon particles of different sizes. Food animals would also be killed in the same catastrophe.

---

[7] In 2004 a Yakuza-generated and Yakuza-steered hurricane passing through the Gulf of Mexico also was accompanied by just such freak waves, which were responsible for much more damage being done to the Gulf oil rigs and pipelines than initially thought. Pipelines on the seafloor were damaged particularly heavily, requiring very costly repairs.

Even worse damage occurred some 74,000 years ago or so, when another large supervolcano exploded, killing all human life on earth except for a few thousand in Africa. More on that shortly.

Most of all, in 1997 Secretary of Defense William Cohen *specifically confirmed* that terrorists already were using electromagnetic weapons to stimulate volcanoes into eruption {xxxv}. Since that is confirmed, and since the Yakuza has the scalar interferometers to do so, it follows that the Yakuza also will have identified the volcanoes in which the most damaging eruptions could be induced to strategically destroy or severely maul the United States. *And so it has.*

To further illustrate the killing power of a supervolcano, consider the eruption, about eleven million years ago, at Bruneau Jarbridge south of Hagerman, Idaho. This eruption covered half of North America with ash two meters thick. Distinctive fossils as far away as Nebraska, 1600 to 2000 kilometers distant, reveal ash piled two meters high that choked the vast animal population of North America to death. Marie's disease, a deadly lung disease caused by ash exposure, contributed to the mass extinction of North American megafauna.

Also consider what probably happened about 74,000 years ago with the explosion of the supervolcano whose caldera today lies under Lake Toba on the Indonesian island of Sumatra. Though the findings are still controversial, increasingly it appears from multiple evidence that the eruption of Toba wiped out all human life on earth except for a few thousand humans in a location in Africa. This follows a formal theory by Professor Stanley H. Ambrose {lv}, which has gained support as a result of independent geological {lvi} and genetic evidence {lvii}.

Genetically, the human mitochondrial genetic trace clearly shows a dramatic and sudden reduction in genetic variability (called a *genetic bottleneck*) some 74,000 years ago. Also, present human genetics shows that the ancestors of every person living today came from Africa. After a period of continuing extinction, then during the last 74,000 years following that violent eruption, every human born on the planet descended from the few thousand humans in Africa who survived. *Apparently all other humans on earth perished, along with their inherited genetic variations, producing a profound genetic bottleneck in the entire human species* {lviii}.

Quoting Professor Lynn Jorde, a specialist at the University of Utah who uses deep analysis of mitochondrial DNA to investigate mankind's past {lvii}:

> "We have a dramatic reduction in genetic diversity during this time when the population is very small and then after the bottleneck the people who we see today would be descendants only of those who survived. And they're going to be genetically much more similar to one another, reducing the amount of genetic variation. ...Mutations in the mitochondria take place with clocklike regularity, so the number of mutations gives us a clock essentially that we can use to approximately date the major event. In the case of a population bottleneck, we think that this would have occurred roughly 70-80,000 years ago, give or take some number of thousands of years. So then the real question

> is: What could have caused such a reduction — an extreme reduction — in the human population down to as few as 5 or 10,000 individuals?"

Increasingly the cause appears to be almost certainly the recognized supervolcano eruption at Toba, some 74,000 years or so ago {lix}.

As another example, there is a volcano named Cumbre Vieja, on the Island of La Palma in the Canary Islands, which split in the past. The split-off western flank – about the size of the entire Isle of Mann – still hangs above the ocean, poised to slip into the sea. Its eventual crash into the sea will evoke massive tsunamis, moving outward at some 600 miles per hour. Within 7 to 10 hours, an immense wall of water, up to several tens of meters high, will strike New York, Boston, and the eastern seaboard, penetrating inland and devastating the area {lx}. It is not a matter of *if* the incident will occur, but only a matter of *when*.

Tsunamis from underwater volcanoes, seafloor earthquakes, and other shocks occur around the world, including in the Pacific Ocean. Quoting {lxi}:

> "In the last 1,000 years, Pacific Ocean tsunamis have been observed and recorded over 1,000 times. All of these were major catastrophic events, since smaller events often go unrecorded."

Cumbre Vieja presents another wonderful opportunity to the Yakuza, who almost certainly has registered their scalar interferometers on it, intending to evoke a tsunami by triggering the volcano into eruption. If so, then this is one of the earthquake eruptions that are almost certain to be initiated by the Yakuza, *during the next two years*, to assist in causing huge economic damage and mass casualties to the east coast of the United States. Incredibly, the U.S. apparently does not monitor Cumbre Vieja to enable hours of advance warning of the giant tsunami. Without a warning system and lead time to evacuate major cities, great loss of life and economic destruction can be expected.

Further, there are conditions off the west coast of the U.S. that are also suitable for induction of undersea quakes and large tsunamis. Undoubtedly these are already on the Yakuza's target list and schedule for striking.

As pointed out by Stockton {lxii}, there is also the possibility of heating methane hydrates (frozen natural gas) on the eastern continental shelf to cause massive undersea landslides, which would also cause tsunamis.

The strategic significance of Secretary of Defense Cohen's statement {xxxv} in April 1997 becomes abundantly clear. *He was referring to a grave strategic destructive capability, already in the hands of the terrorists and available. He was confirming that this capability is already being used by the terrorists in their asymmetric war against us.*

That strategic destructive capability, when adroitly utilized, as in the simple examples given, is as powerful as the full nuclear ICBM capability of the Soviet Union or the United States.

Since the Yakuza interferometers can also massively intervene in submarine volcanic and earthquake activity on the ocean floor worldwide, their capability to induce giant tsunamis and other violent waves, in addition to giant quakes and volcanic eruptions, is of the gravest significance.

Indeed, the Yakuza recently spectacularly demonstrated such effects, under the direction of the FSB/KGB. On Dec. 26, 2004 they initiated the magnitude 9.3 seafloor quake that caused severe tsunamis and loss of life in Asia, now thought to include more than 300,000 deaths in addition to enormous economic damage.

Summary of Some Weather Engineering and Other Capabilities

Some of the simpler scalar interferometer capabilities in engineering of the weather and geophysical events are:

a. Steering the jet stream, thus steering a weather entity (i.e., storm, front, air mass, and hurricane). Warming the air in one region expand it so it becomes less dense, creating a low pressure area. Gradually moving the target interference zone determines its path. Cooling the air in a region produces the opposite effect, a high pressure zone, steerable in the same fashion. By making multiple highs and lows and adroitly positioning and steering them, the jet streams and other prevailing winds can be entrained, captured and steered. This alone allows substantial augmentation and steering of weather effects.

b. Warming or cooling the atmosphere in a storm or front to affect its power.

c. Producing sharp *negative energy* EM pulses inside a storm so that the negative energy Dirac sea holes "eat" electron charges, reducing the storm's charge and power.

d. Producing *positive energy* EM pulses just outside a storm in order to lift extra electrons from the Dirac Sea, to increase the storm's charge and power. In that case the resulting Dirac holes expand outward and away from the storm, having no effect on it. As they meet stray electrons, these holes eat the electrons and both disappear back into the vacuum, *as* ordinary vacuum (the Dirac Sea).

e. Building very large rotational atmospheric flow in the path of a storm, and then sharply decreasing the diameter of its curvature, so that the storm's angular momentum spawns concentrated spin flows or meteorological circulations, to form tornadoes, waterspouts, etc.

f. Establishing a large high pressure area in a region to block lower-pressure normal winds and storm directions.

g. Establishing a large low pressure area in a region to attract higher-pressure normal winds and storm activity.

h. Focusing the interference zone inside a volcano's magma chamber and steadily depositing additional EM energy there builds up internal pressure, causing an eruption. A slow rise in pressure, resisted by friction, engenders a large and violent eruption, and consequently maximizes the distant dispersion of great quantities of ash, lava, and other debris.

i. By more rapidly inputting the energy into the volcano, the eruption will occur at a lower pressure and so a smaller eruption will ensue.

j. All sorts of EM energy forms — glowing spheres, hemispheres, etc. — can be produced and used against various targets, for electrical and electronic destruction or for an electromagnetic explosion when the energy contacts the intended target.

k. Depositing extra interferometry energy into a fault zone increases stress and piezoelectric activity in the rocks, thereby inducing an earthquake when the rocks finally slip. By adjusting the rate at which the excess energy is added, the size of the resulting earthquake can also be changed.

l. By increasing the charge of a storm, its electrical activity and the resulting lightning strikes can be increased. This is useful, e.g., in repeatedly producing large forest fires, particularly when combined with weather engineering to produce a drought condition. On the other hand, by decreasing the charge of a storm, its electrical activity and therefore its lightning strike activity can be decreased.

m. By inducing repeatedly pulsed negative energy in an area containing living animals or humans, the animals or humans can be directly killed. Strong pulsing will result in rather instant death, where the bodies drop limply, with not even a nerve cell firing thereafter. Everything living — cells, microbes, viruses, whatever — in the powerfully struck bodies is killed instantly, and the bodies do not decay, even over a month or more. The Soviets tested such weaponry in this mode against two Afghan villages in their own war in Afghanistan, and it is probable that the Yakuza has produced portable negative energy EMP weapons with similar capability.

n. The longitudinal EM wave sets — and beams of such — transmitted by a scalar interferometer easily penetrate Faraday shields, the ocean, the earth, etc. That is because matter is mostly empty space filled with Whittaker EM fields and potentials, which in turn are bundles of longitudinal EM wave sets. Hence matter is a "superhighway" for the throughput of longitudinal waves. Such longitudinal wave weapons are very useful in attacking deep underground targets and facilities and destroying them or wreaking damage. As an example, the leader of a nation can be killed sitting in the chair at his desk, by a precision scalar interferometer. The interferometry focus has been demonstrated down to a few inches. In Japan, the Yakuza has demonstrated this capability on a few occasions to kill several Japanese critics inside shelters or shielded facilities. The significant impact on the very notion of shielding by strategic command and control centers in underground bunkers and facilities can be easily appreciated. Essentially all facilities and personnel become completely vulnerable.

o. The use of a more portable scalar interferometer to destroy the electronic controls of a normal or nuclear power plant from a distance is obvious. With nuclear power plants this poses the risk of a melt down. Electrical controls for pipe valves, etc. are also vulnerable; spent nuclear fuel rods are largely stored in underwater pools on normal power plant sites. If valves open and the water drains from those pools, the rods will heat up and a melt down or very hazardous venting of radioactivity can ensue.

p. The control systems for hydroelectric dams are deadly vulnerable to scalar interferometry attack.

q. Large electronic complexes such as switching and control systems, centralized control systems for power grids and substations, etc. are highly vulnerable to attack by long- or short-range scalar interferometers.

r. Chemical plants, refineries, fuel storage sites, tank farms, etc. are also highly vulnerable to scalar interferometry, including to portable attack.

s. Liquid natural gas ships, oil tankers, etc. are also extremely vulnerable to scalar interferometry attack.

t. Surges on the power distribution grid transmission lines can easily be accomplished by scalar interferometry at will, methodically causing great damage to the grid and emergency shutdown of most of its "feeding" power plants. As much of the grid as desired can easily be kept nonfunctional, as long as desired, by occasional repeats from a single long-range scalar interferometer.

u. Any communications center or headquarters is completely vulnerable to attack and destruction by strategic or portable scalar interferometers. Since a scalar interferometer can also produce *negative energy* EMP pulses of extreme lethality in the interference zone, all centralized command and control systems are essentially totally vulnerable. As an example, our own future network centric warfare communications systems can be totally destroyed in a few minutes by only a few scalar interferometers, even by just one.

v. Many other capabilities also exist to use the interferometry to affect the weather, the atmosphere, the rocks in the earth or at the bottom of the ocean, etc.

Once the Yakuza scalar interferometers are factored into the strategic terrorist threat situation, there is no recourse that can save the present centralized electrical power system, large U.S. command and control systems, etc.

The only way to prevent the total economic collapse of the United States induced by collapsing the centralized electrical power system is to immediately initiate the strongest possible energy from the vacuum systems development and production program. For the nation to survive the coming destruction of its power system, the goal must be to replace the entire centralized power system, including its long distribution lines and large feeder power plants, with decentralized EFTV power systems as rapidly as possible. That centralized system is simply not defensible and not sustainable against current threats, and against the hostile assets already successfully inserted or waiting abroad to support terrorist operations inside the U.S.

So the strongest possible immediate efforts are necessary if we are to prevent our total economic collapse in about two years, and also our subsequent total destruction by combined scalar interferometry and internal terrorist attacks. A massive Manhattan-type project to develop and deploy EFTV electrical power systems is most urgently required, as at least equal to the highest U.S. priority. We bluntly state without further discussion that such EFTV systems are indeed already developed in our own country and worldwide, and so they are available, but they have deliberately been withheld or suppressed for decades. Now they must be unleashed.

Short of a direct Presidential order and Presidential Decision Directive, there appears to be no way that the necessary program can be or will be galvanized, because every powerful scientific, industrial, and government laboratory cartel in our society is bitterly opposed to any such program or action. Our morose assessment is that, *unless some other major strategic factor significantly enters the conflict*, the Yakuza and other terrorists — attacking our present centralized and totally vulnerable control, planning, and energy systems — will probably succeed in catastrophically destroying the U.S. economy on schedule. If so, that will leave the U.S. prostrate for its subsequent rapid physical destruction by the jubilant Yakuza in a giant "turkey shoot". And hence our oblivion.

U.S. Societal Polarizations Are Being Gradually Increased

Among sociologists, there is much support for the notion that societal polarizations naturally tend to increase {lxiii}. Further, many pollsters state that the nation is very seriously polarized already, particularly since the recent presidential election. For example, quoting John Zogby on the 2004 elections {lxiv}:

> "I have been calling this the Armageddon election now for months because we are so polarized, so split culturally, politically, ideologically, demographically, like almost no other time in American history. ... The last time the nation was this deeply divided over what course to take was the original Armageddon election of 1800."

However, for the first time in history, certain rather esoteric "mind war" FSB/KGB weapons are being directly employed to gradually and reliably intensify the societal polarizations in America — straight vs. gay, conservative vs. liberal, Republican vs. Democrat, one religion against another, one race against another, labor class vs. financial class, etc. The strategic objective is to increase and intensify the extreme ends of these polarizations, by altering the behavior of certain leaders. With intensification, the polarizations will slowly become so fierce that, about two years from now, serious rioting and bloodshed massively spilling into the streets and cities will result in chaos, martial law, etc. and severe polarization and disorganization. In short, the society will be engaged in fiercely consuming itself.

So far, the deliberate escalation of our societal polarizations by the new mind war and conversion techniques is right on schedule, as anyone can see now, and will increasingly see over the next two years.

U.S. societal polarizations will continue to increase now that the election is over. Temporary *conversions* and *fractional conversions* {vii} of the minds and mind operations (and therefore the behavior) of selected leaders or prime movers in the various polarization areas are being used and will increase. This capability arises from the branch of energetics known as *psychoenergetics*, and such measures are now being used selectively to intensify our polarizations. Particularly see our giant briefing on mind war and conversions, in the second part of this book.

Dramatic tests of the distant, total control over the mind and behavior of a person performing an intensive technical task (flying and operating an aircraft as a weapons platform) {lxv} were conducted against Captains Button and Svoboda in 1997. A third test against Captain Hess in 1998 demonstrated total control of the autonomic nervous system as well. Those dramatic "suicides" were deliberate, high profile stimuli to see if our own intelligence analysts knew anything about psychoenergetics yet. The answer was a very resounding and clear *"No, they are still ignorant of that area, they are very determined to remain ignorant of that area, and hence the nation is still deadly vulnerable to use of such weapons and effects."*

Our scientific community still does not understand that mind operations are totally electromagnetic but temporal, not spatial, although they also have *virtual state* projections in 3-space. Because the scientific community has suppressed the source charge problem, they have no notion of the exact mechanism by which virtual state changes can be integrated to observable changes, electromagnetically by the action of charge itself.

They also do not understand that observations in 3-space have small virtual state projections in the time domain (as mind operations). Our scientists have no understanding at all that mind operations are (i) electromagnetic, (ii) temporal, and (iii) directly engineerable and creatable (but using time-polarized EM photons and waves). They do not understand what thought is, what intent is and what its exact mechanism of expression is, and how mind and body are coupled to provide a living biological system.

Again, because our nation is so totally vulnerable to mind war conversions and fractional conversions, a U.S. crash program of utmost rapidity to catch up in that area, under a Presidential Decision Directive, is urgently required.

For example, on his customary morning exercise jog at a military post in the U.S., in March 1998 Captain Gordon Hess suddenly veered aside, sat down, took out his knife and calmly stabbed himself in the neck and torso some 26 times, including multiple thrusts directly through the heart. Note that the shock of a stab through the heart is considered in forensic medicine as generating such shock and recoil of the autonomic nervous system that it is not possible for an intentional suicide to continue, much less to achieve a total of 26 repeated stabs until the body simply became nonfunctional and toppled over.

In short, total control of even the deep autonomic nervous system was demonstrated resoundingly. The purpose of the test was two-fold: (i) demonstrate distant instant conversion and total control, including of the autonomic nervous system itself, and (ii) stimulate and "ping" U.S. intelligence to see if they knew what was going on, what had been tested, and why. Both purposes were achieved. The test went perfectly, and the U.S. demonstrated it was totally asleep and totally ignorant of such technology, and even of the technology's *ansatz*. Indeed, by order of the U.S. Army, two days later front loaders dumped two tons of dirt on the death scene, effectively hiding any further evidence. This is reminiscent of a similar U.S. government action at the crash site of the Arrow DC-8 at Gander Air Force Base, Newfoundland in 1985. A general officer arrived very quickly on the scene, and had the entire area bulldozed after extracting the wreckage.

As stated, the Yakuza is a major arm and the protégé of the FSB/KGB's coordinated asymmetric war and terrorism, carrying the bulk of the strategic war to be conducted within the U.S., and reporting to those die-hard old communists who are the hidden FSB/KGB directors. The Yakuza view is that — together with the other terrorist organizations already having infiltrated substantial assets into this nation, and together with chaos in U.S. streets due to severe polarizations gradually being intensified now — they are already helping to bleed the dragon increasingly. Then as the scenario progresses they will joyfully begin the direct destruction of the centralized U.S. electrical power system and other key economic targets (harbors, refineries, pipelines, etc.). They will also participate in immune system spreading (explained later in this paper), as a force amplifier to biological warfare strikes by other terrorists. As all of this hostile action and destruction ramps up over the next two years, it will crescendo and catastrophically collapse the entire economy of the U.S.

Added to assure that great economic collapse, there will also be a sudden and massive collapse of "group ordering and coherence" (GO) induced by massive conversions of "sleeper" inserted conversions and fractional conversions. E.g., by sudden switching conversion of massive numbers of key persons whose behavior instantly and dramatically changes all over the map, the order and coherence of any important group can be and will be devastated and lost, immediately. Our police forces may suddenly begin attacking our populace, while crews manning and operating our ICBMs, nuclear subs, and nuclear bombers "go crazy" and erratically start firing at our own cities, and at each other. And so on, in a society that has suddenly erupted into total schizophrenia. Throughout the U.S. and Canada, GO will rather dramatically become NO-GO, as all group coherence is lost and total chaos sets in, as in the two examples given..

Given the planned present offensive schedule and the insertions already accomplished, indeed the plan for full strategic destruction of the U.S. is likely to succeed.

The psychoenergetics conversions and fractional conversions are a major strategic weapons program in support of the present three-year "Operations Phase" schedule for strategic destruction of the U.S. So it is also a program that supports the strategic destruction of the centralized U.S. electrical power system including its infrastructure, as one purpose.

The *original* targets desired by Bin Laden for the 9/11 attack actually were nuclear power plants etc. — *parts of the centralized electrical power system* — rather than the more symbolic Twin Towers, Pentagon, and White House. The targeting was apparently changed by Bin Laden himself. He felt that his assets in place and available were not yet sufficiently trained to produce a truly effective, coordinated simultaneous set of attacks on nuclear power plants and other "harder" targets. So to guarantee success, he opted for strikes on symbolic but softer targets, which would provide great publicity stimulus in the MidEast for his cause, in addition to causing great consternation in the United States.

Insertions Far Exceeding Public Acknowledgment

Long before its economic collapse in the latter 1980s, the former Soviet Union inserted nuclear weapons in all our major cities and population centers, along with Spetznaz teams to detonate them on command. Lunev's book {lxvi} even tells a few of the ways the weapons were inserted. Those weapons and teams, still controlled by the old-guard Communist faction that dominates the FSB/KGB, are still here, still waiting for the order to detonate the nuclear weapons and destroy the United States. The Communist faction (not the younger and more liberal faction from which Putin comes) is the one plotting to destroy the United States.

These Russian nuclear weapons and teams may be brought into play during the operations phase as several substantial nuclear explosions in major U.S. population centers that will be blamed on the international terrorists. The result will be great loss of life, nuclear contamination of several large U.S. cities, complete overwhelming of our Homeland Defense and medical capability to handle mass casualties, and terrible physical damage to further guarantee the scheduled catastrophic U.S. economic collapse once the centralized electrical power system is taken out.

As previously mentioned, we would have been destroyed by the former Soviet Union in the late 1970s, by those successfully inserted Soviet nuclear weapons already at their target sites on station in the U.S., except that a friendly nation also inserted nuclear weapons massively into the cities and key target areas of Russia. Those friendly weapons are also still in place, waiting for the order for their detonation if necessary. This sort of thing in the "old days" of cold war intelligence was called *dead man fuzing*. It is still in place today, but its "range of uses and functions" has substantially increased.

Being chess players and *rational adversaries*, the Russians can be countered by a balance of terror including dead man fuzing — the old Mutual Assured Destruction (MAD) notion with the twist of using inserted nuclear weapons already on site in the targeted nation — because they will not just commit mass suicide for Mother Russia. So the United States survived in the 1970s, courtesy of a friendly nation, and we *were not* suddenly destroyed by American city after city blowing up in a nuclear inferno. The MAD tactic worked because the adversaries were known and their location could be destroyed at will.

However, with irrational foes such as those in the war on Islamic terrorism, mutual assured destruction is not an effective deterrent. There is no single nation or location in which to insert "dead man fuzing" against the terrorists or target with deliverable weapons of mass destruction. The present terrorists are not localized, but are dispersed worldwide and embedded in the civilian populaces of many nations, including our own. In that case, *there is only assured destruction of the nation targeted by the terrorists, if and when the terrorists successfully place enough weapons of mass destruction into the targeted nation(s). In short, MAD simply turns into AD. The terrorists will soon achieve that capability, if they have not already reached it.*

In the case of assured destruction, the insertion phase — which can reach a very high assurance level before the operations phase is initiated — becomes the key to winning or losing, given that the terrorists have or obtain sufficient WMD assets to insert.

The purpose of strategic asymmetric warfare is to kill or destroy the targeted populace. It is a fanatical ultimate expression of General George Patton's famous, blunt statement:

> "I want you to remember that no bastard ever won a war by dying for his country. He won it by making the other poor dumb bastard die for his country..." {lxvii}.

Modern Asymmetrical War: The Scorpion Duel

We thus are back to the absolutely primitive warfare of two dueling scorpions: He who first succeeds at insertion, and then strikes his enemy hard and decisively in his vitals, is he who wins a priori.

The procedure is to first ensure that one will hit the foe's vitals catastrophically, and sufficiently to destroy the entire foe himself. That is the *Insertion Phase*. Then the procedure is to unleash what has been assembled, thereby destroying the foe. That is the *Operations Phase*.

Hence the success of the *insertion phase* — inserting the necessary WMD and operational teams into those vitals in advance — predetermines the success of the subsequent *operations phase*, where one strikes on order, unleashing the WMD and destroying the nation's vitals and therefore destroying the foe.

For the targeted scorpion, there is no turning back or delay, once the insertion phase of the scorpion duel begins against him. *Successful insertion of the required WMD and WMD teams — during a time our folks innocently think of as "peacetime" — almost inevitably wins the war by destroying the penetrated adversary very shortly after the operations phase begins*. In a nutshell, those two phases — insertion followed by operations — constitute the "offense" part of asymmetric war. And since peacetime is just the insertion phase, *asymmetric war is perpetual war, whether or not shooting is ongoing*.

There has been little public "briefing" by our government on what the estimated success of the insertion phase against us has been and is now. Apparently much of the analysis is in disarray, filled with assumptions, and very short on actual definitive data. It also follows that most definitive data in this area would be highly classified, particularly to protect effective human intelligence (HUMINT) sources.

Now one can appreciate VP Cheney's morose statement. Since "peacetime" now is just another name for "insertion phase", then "perpetual peace" is still "perpetual war" in the new vernacular. So we have entered a "perpetual war" we have not previously known or understood. There is no real peace in the sense we think of it. *There never will be any peace again, until one opposing side of the dueling scorpions is utterly destroyed.*

The "defense" part for the terrorists is to hide or defend their own "vitals" — their actual bases, support organizations, supply routes, troops and forces and equipment, etc. In the present asymmetric war, there are no single terrorist bases for us to just "knock out" decisively. We cannot win this war by simply having our troops riding out of "Fortress America", meeting and defending some hostile army on some distant battlefield, and then riding back into our protected fortress again. *There no longer exists such a thing as a "protected Fortress America".*

Instead of an identifiable single army on a single foreign battlefield, the terrorist foe's "vitals" are hidden around the world, in many places, and in many ways. Many of them — often the most dangerous of them — are secretly hidden right here in the United States, inserted during the insertion phase which has been ongoing for at least 20 years.

In the U.S., we cannot "hide" our vitals such as refineries, pipelines, railroads, bridges, electrical power plants and long transmission lines, harbors, water supplies, food supplies, etc. Instead, we can only try to protect them against their certain attack, and try to limit the damage and destruction. *The main target for the terrorists is not just our military forces, but is primarily our civilian populace and especially our national economy — which is particularly vulnerable by the rather simple destruction of our centralized electrical power system and its primary support infrastructure.*

The real art of asymmetric war is to first draw most of the targeted opponent's military forces out of the protected fortress and get them scattered widely in distant locations and distant struggles. By augmenting the distant resistance, this forces the targeted nation's military forces mostly to be deployed elsewhere, leaving the internal fortress itself much less protected, and thus making it highly vulnerable.

Once the major decisive action is shifted to inside the protected fortress, there is an urgent necessity for (i) a strong Homeland Defense, (ii) well-organized and *effective* mass casualty treatment capabilities (which presently do not exist), and (iii) strong border defenses, sealing of the borders, and interception of insertion attempts as they occur, etc. This also presently is not in place; our borders are full of holes and in many ways almost a joke. There is also the urgent need to go after the opposing scorpion's "vitals" wherever the pieces can be located and "fixed in place" and destroyed.

Ironically, for the U.S. these urgent needs come at a time when our HUMINT resources have largely been abandoned or nullified, by the past fascination of our intelligence community for technical widgets such as satellite photography, etc. So our ability to ferret out inserted cells and WMD has been sharply reduced by the intelligence community itself.

Effective HUMINT takes years to get established with successful agents infiltrated into the terrorist cells and planning organizations. So not too much is going to be rapidly accomplished to resolve our HUMINT shortfalls. Much of what HUMINT is available is either from foreign sources or is due to the clandestine cooperation of much of the peaceful U.S. Muslim community, which sees this nation as its own country, and thus sees any threat to destroy the U.S. as a threat against the peaceful Muslim religion itself. To understand some of the factions and divisions of the Muslim world, the reader is referred to treatises by Williams {xcvi, xcvii}.

But the scorpion duel is highly sensitive to being able to (i) find the opponents vitals (which now are the inserted assets and terrorist teams within one's own nation) and (ii) go after the scorpion's vitals and destroy them in what are erroneously called "peacetime operations". Ironically, one sets up winning or losing the asymmetric war, by one's actions in peacetime. *Once in an asymmetric war, "peacetime" is the absence of significant* ***overt*** *active hostilities, but with the presence of fierce* ***covert*** *insertion and preparations for the follow-on operations phase and the "knockout" punches.*

Distant Military Actions: Necessary but Not Sufficient

Hence one cannot think of winning major battles and the asymmetric war in a few months or even a few years, simply by fighting successfully on well-defined distant battlefields. Those distant conflicts and confrontations are *necessary but not sufficient*, a term mathematicians delight in using. It simply means one has to do it, to prevent losing, but by itself it is not sufficient to win the war on terrorism.

Instead, one must also think of a continuing gritty, gruesome war, lasting for decades, ongoing in one's own cities and population centers and installations. One rather gradually grinds down the terrorist opponents and rather gradually uncovers and destroys the various parts of their vitals, while protecting one's own vitals as much as possible, and while rooting out everything one can of the insertions made by the foe or further attempted by the foe. *Here effective HUMINT is vitally important, and our shortage of it is a grave strategic vulnerability.* Obviously our intelligence agencies are also scrambling to try to get such HUMINT effectively going again.

If one destroys the terrorists' assets and capabilities before they destroy him, one survives, but still must root and destroy the terrorists themselves. If one does not succeed in doing that, one dies because the terrorists will resurge and destroy him without mercy. *This is the message the Al Qaida and other terrorists are symbolically sending, when they behead prisoners and hostages.* It is absolutely a scorpion duel to the death.

With the abrupt wake-up call of 9/11, we were suddenly confronted with the continuing and unrelenting nature of this asymmetric warfare. At least officialdom and our armed forces were confronted with it; there has been no truly adequate news briefing or educational series for the American public on such matters.

The military schools of the Army, Navy, and Air Force do include education on such asymmetric warfare, including highly classified courses at the top secret level. But certainly the U.S. public still does not understand the absolute necessity for Cheney's morose statement or the full nature of the new reality of asymmetric war {xlviii}. Indeed, much of our population is polarized, because of — for one thing — a substantial and naïve belief that all problems could readily be solved by negotiation, if only Bush were so inclined.

Indeed, as with every other strong and emotional subject, the subject of "asymmetry" in warfare is itself a societal polarization area. Typical "anti-views" simply downplay the very notion of asymmetry as being "nothing new or special" in warfare {lxviii}. As one reads such treatises, it becomes crystal clear that our own strategic planners and strategic theorists themselves are often confused, simply recasting asymmetric war into what has gone before. They are not factoring in such radical developments as the Yakuza's acquisition and use of scalar electromagnetic superweapons, or the dramatic change that occurs in the theory once full strategic U.S. destruction is a capability already in terrorists' hands, with multiple knock-out blows already cocked and waiting. And they certainly are not factoring in the stunning implications of mind war, conversions, etc.

If one can disable an entire populace mentally, at a single stroke, one can use such "mind war" conversion switching to destroy any order in the society nearly instantly.

With suitable preparation during the insertion phase, terrorists and foes with such capability can indeed achieve sudden massive "knockout" of a targeted populace, in ways presently not dreamed of by the open press.

Asymmetric Warfare Simplified: An Overview

Let us briefly discuss this "new" kind of "old" warfare in the most direct manner possible. The present war is a very different kind of warfare called ***asymmetric*** war. It consists of two phases: (1) the ***insertion phase***, (conducted in what we erroneously call "peacetime"), during which the necessary WMD and teams to operate and unleash them are inserted in the targeted nation's populace, population centers, and key target areas and made ready to go, and (2) the ***operations phase***, in which the inserted forces and assets are unleashed, thereby destroying the targeted nation — in this case, the U.S. itself. The bulk of the target and intention is to kill U.S. civilians and destroy critical facilities right in the U.S. homeland — as many as possible, as efficiently as possible, and wherever and whenever possible.

Secondary[8] to the importance of population kill during the operations phase is the destruction of the targeted nation's economy, catastrophically and decisively. If the foe succeeds dramatically in accomplishing the *insertion phase*, the foe will almost automatically win the follow-on *operations phase* a priori, unless he is an utter fool.

Further, the development of secret weapons and superweapons is not confined just to their use in overt warfare. Depending on their nature, they can also be used in covert or asymmetric warfare, in great surprise and shock upon the targeted populace.

Our very real problem today is that the insertion phase has been going on for at least two decades, while we were literally "sleeping" as far as intercepting and degrading the insertion effort by any substantial degree at all. *The blunt truth is that the terrorists' Insertion Phase has just about been successfully accomplished already. We are in grave danger of being utterly destroyed, with high assurance, in the three-year Operations Phase whose first part has already begun.*

Compare the conduct of asymmetric warfare as analogous to orienting and setting up weapons and ammunition and target lists for a massive artillery barrage, and preparing to make a massive attack with that barrage. One first gets the artillery guns shipped into the necessary assembly areas, assembled in place, with their ammunition and crews inserted and made ready, with their target lists completed and waiting. Then one gets the vast assembly of guns and crews properly *registered* on their specific targets and ready to go. That's the "insertion" or "preparation" phase for a normal mass artillery bombardment.

Then the assemblage of "weapons in place" simply fires the massive barrage on order, and that is the "operations" phase, exploding with suddenness and great violence right on the enemy's vitals. An important major characteristic of this "normal" warfare insertion phase is that it is the insertion of the delivery systems, to chosen "attack positions" *outside* the targeted area.

In asymmetric war, instead of assembling all the external delivery systems (cannon, rockets, missiles, bombers, etc.) it is as if *most of the deadly ordnance is already delivered and is assembled right in its target areas, inside the targeted nation, along with the necessary teams to set them off right there in place.* Long-range additional systems placed "outside" the targeted nations can also assist or lend support, particularly if the systems are still secret and unknown to the targeted nation.

---

[8] However, if the economy can be catastrophically destroyed along with total disordering of the society by sudden massive mind conversions, destroying all coherence, the entire population is reduced to hapless targets and already completely defeated. Such a hapless populace can then be methodically destroyed.

In short, in asymmetric warfare the "delivery" part of the former operations phase is pre-accomplished, before hostilities even begin. So the insertion phase gets the artillery rounds placed right at the target during "peacetime", and avoids any further assembling of the external artillery pieces, rocket launchers, and other delivery systems necessary in the conventional war approach.

Thus *successful insertion* in asymmetric war overcomes one of the vagaries of standard war: the inaccuracies and mishaps caused by misdeliveries and/or losses of the delivery systems themselves prior to delivering the ordnance on the distant targets. In normal war, those misdeliveries and losses to the delivery systems directly subtract from the effects achieved on the target once the attack is initiated. In *asymmetric* war, the cost of misdeliveries and losses to the inserted WMD are paid for before the hostilities and before the attack (the operations phase) begins. So given a successful insertion, "delivery system losses" prior to attack do not subtract from the effectiveness of the attack results, once the attack is initiated. Indeed, successful insertion is a great *force amplifier*.

Once the hostile terrorists possess weapons of mass destruction and have already delivered them onsite to the actual targets, the "massive barrage" of the operations phase is certain to destroy the targeted foe, if the terrorist leaders have any competence at all in their planning and preparations. *The great advantage of the insertion phase is that it covertly prepares a great "barrage" where the "ordnance delivery to the exact target locations" has already been accomplished and its lethal effectiveness against the actual targets has therefore been assured in advance.* With proper stealth and intrigue, the insertion phase can be accomplished slowly and at will, and without any extraordinary resistance being offered by the "sleeping" targeted nation or population! After all, it is "peacetime" when such clandestine insertion operations are conducted. *In the new concept, "peace" is simply the insertion phase of the perpetual war.*

So asymmetric war bypasses much of the normal overt concern with long range delivery systems such as ICBMs, strategic bombers, aircraft carriers, nuclear submarines, etc. The usual giant armada of *delivery systems* is largely unnecessary, if the terrorists can successfully perform the covert insertion phase. Instead, the terrorists' major early activity is in ordnance procurement, not in building long range delivery systems.

Obviously asymmetric war dramatically differs dramatically from normal symmetric war, and so there are definitely some new questions which must be asked and answered.

A Few Questions to be Answered

In an asymmetric war situation, some major appropriate questions for the targeted nation are: Are all the terrorists identified as to parent organization, etc.? How certain is the targeted nation that there have been no substantial omissions from its foe identification and capabilities list? Are there already effective compensatory plans in case there suddenly are uncovered additional terrorist organizations or parent organizations, and/or additional capabilities?

What is the WMD and other clandestine weapon capability of the terrorists, their organizations, their supporting nations or groups, etc.? How certain is the targeted nation that there have been no substantial omissions from the weapons capability lists? Are there already effective compensatory plans in case there suddenly are uncovered additional WMD or clandestine weapon capabilities of the terrorists or their parent organizations?

In view of Secretary of Defense William Cohen's seminal 1997 statement, what specifically is the capability of the terrorists to cause those volcanic eruptions, weather engineering and climate control, and initiation of earthquakes? Particularly in view of anomalous quake swarms ongoing, and the disastrous 9.3 undersea quake on Dec. 26, 2004 that initiated the catastrophic tsunamis in Asia? What specifically is the nature of these strange "electromagnetic" weapons confirmed by Defense Secretary Cohen? What specifically are the counters to them, and specifically how effective are these counters? Further, why is so large a percentage of the government still unknowledgeable of this critical portion of the terrorist threat and thus totally unprepared for it? Why has the American public not been completely and adequately briefed on these *confirmed* terrorist developments?

Even though known specific foreign areas with concentrations of the terrorist organizations and/or their support groups are targeted for attack and neutralization, how much of the total terrorist capabilities have been reduced by such measures? What if anything has been the impact on the inserted capabilities already inside the United States?

Exactly what information is officially available (and to what degree of confidence) on the "missing" 5 Ukraine SS-19 Stiletto ICBMs, each missile with 6 nuclear warheads and each warhead of several hundred kilotons yield? {lxix} To add spice to the controversy, three years after "all" the SS-19s were supposedly turned in and accounted for, another 31 *physically* turned up in Ukrainian silos and warehouses! One also questions what happened to the additional 230 or so missing "extra" nuclear warheads from the Ukraine total? {lxx, lxxi} What happened to the missing Russian suitcase nuclear weapons? {lxxii} What happened to the 20 suitcase nuclear weapons allegedly purchased by Bin Laden {lxxiii}, etc.? What if anything has been done about the already-inserted Russian nuclear weapons and Spetznaz teams in our cities and population centers? What if any defense or countermeasures exist to prevent our destruction at some point, from these nuclear weapons alone? And again, why has not the American public been adequately briefed on such hostile capabilities and on the "nuclear knives already at our throats"?

How long has the insertion phase been ongoing? How effective or ineffective has interception and neutralization or intercept of the insertion attempts been? Again, why has the American public not been thoroughly briefed on these areas? Why has the Congress not been adequately briefed on them? What is Homeland Defense doing about them?

What is the probable target listing believed to have been compiled by the terrorists, based on their actual weapon capabilities and the degree of insertion success actually accomplished? Considering Bin Laden's background in economics, and his past statements on the importance of disrupting the economics, is not the near-absolute vulnerability of our centralized electrical power system our greatest strategic weakness, insofar as catastrophic collapse of the U.S. economy being introduced? If so, why has the scientific community and industrial community been allowed to push for hot fusion, hundreds of pebble bed reactor nuclear power plants, etc. when our entire centralized electrical power system will be destroyed long before any of those could come on line? Even if the nuclear power plants were built and on line, they too are "sitting ducks" and "fat targets", so what is to prevent their easy instruction by the Yakuza scalar interferometers, both long range external systems and portable inserted systems?

What are the various options for the assessed terrorist "preparation" at the particular stage of their insertion phase, or in the near future? Which of these options have the greatest probabilities of destroying or inflicting unacceptable damage on the targeted nation? Which of these options are also matched by known or detected insertions already accomplished or attempted?

Since we know that certain of our enemies — e.g., the KGB/FSB — understand the importance of surprise weapons or superweapons, what is our very best assessment of the probability and capability for the terrorists possessing and inserting surprise weapons or superweapons for which the targeted nation has little defense capabilities? E.g., what about the surprise "electromagnetic" weapons the U.S. Secretary of Defense confirmed, while most of the U.S. government officials — and even our own intelligence agencies, which still largely base their EM intelligence analyses on the totally inadequate electrical engineering model — are still unaware of them? Why has not our scientific community been made fully aware of them, and put to intensive work on them? And since it is perhaps our greatest vulnerability, why is an aged piece of inadequate EM modeling with known serious falsities still perpetuated by the U.S. scientific community, NAS, NSF, NAE, the DoE, etc.? Who's competently watching the "technical store" anyway — is anyone? Or is everyone just blithely accepting the self-serving scientific cartels' pronouncement that "we are the greatest"?

With a three-year plan already in motion for our total destruction, and with a large key role played by the vulnerability of our electrical power system and its coming destruction, why are not our Department of Energy, National Academy of Sciences, National Science Foundation, etc. already under strong Presidential orders to reprioritize their energy research at the greatest speed possible? Why has not the U.S. scientific community already corrected the terribly flawed CEM/EM mess that is taught in our universities and used by our power engineers? Why are they not conducting a massive Manhattan-type project for energy from the vacuum systems to totally replace the present centralized power system? There is no other possible solution, once one recognizes that *the centralized electrical power system itself is going to be destroyed with very high assurance, within about two years from now*. So why are all our scientific efforts actually siphoning off research funds for puerile things which will take decades and will not do the job even then, while enriching and perpetuating the "scientific entourages" driving this sad situation? Since a major part of the problem is the great leap forward of the Russian science of energetics — now 50 years old — and its development of superweapons, then why are our scientists still largely preoccupied with standard electrical engineering and classical electrodynamics — models so hideously flawed and made so antiquated by the progress of physics that they should have been dramatically changed at least three quarters of a century ago?

Why is there no really substantive understanding or definitive work ongoing to rapidly develop means of collecting and using the unlimited free EM energy pouring from every charge and dipole in the universe, including in every EM circuit and self-restricted electrical power system?

Many other similar questions can be and should be raised, but the foregoing questions demonstrate the direction and intent of what should be queried.

The Ease with which Terrorists Could Obtain WMDs

In facing an asymmetric war launched against us, the availability of clandestine WMD to the terrorists, so that the weapons can be inserted, becomes a primary consideration — along with an extraordinary need to understand the new superweapons and rapidly develop counters and defenses.

An internet search for two search terms: Voz Island in the Aral Sea {lxxiv} and anthrax, will quickly demonstrate the ready availability of WMD to any competent and supported terrorist organization. The Russian biological weapons research and production facilities there were the world's best, appreciably ahead of similar facilities in the West. Hundreds of tons of very high quality biological warfare agents were just dumped on Voz Island from those Russian facilities. Quoting Marshall, Malakoff, and Holden {lxxiv}:

> "The United States and Uzbekistan are close to finalizing plans for a $6 million cleanup of a former Soviet bioweapons facility. The effort is aimed at preventing terrorists from harvesting live anthrax spores from a secret dumping ground.
>
> For nearly 60 years starting in the 1930s, the Soviets released anthrax, plague, and other weaponized pathogens on Vozrozhdeniye (Resurrection) Island in the Middle of the Aral Sea. In 1988, at the end of the Cold War, weaponeers buried tons of a particularly potent strain of powdered anthrax at the site, mixing the bacteria with bleach in steel drums to kill it. But several years ago testers found that some of the anthrax is still alive, and water diversions from the shrinking Aral Sea have since opened a land bridge to the once isolated island. Fearing

> that terrorists might try to harvest ready-made bioweapons from the site, U.S. Officials agreed in October to pay for destroying the anthrax and a nearby testing facility."

Actually, for quite some time, any funded terrorist organization that wished a copious supply of the finest anthrax ever made, and wished a strain already resistant to many of our treatments, could just go to Voz Island, pay the natives a little sum, and dig up the dirt in the appropriate part of the dump site, sack it up, and carry it away. Or pay the natives a bit more, and they would dig it up and sack it for the buyer. Wash out the dirt, and one would have all one wanted of the highest grade anthrax spores ever made. *It's safe to assume that those terrorist organizations who wanted a copious supply of high grade anthrax already obtained it with relative ease.*

Bin Laden also obtained anthrax and other BW agents from other sources {lxxv}, as well as at least a few nuclear weapons.

Other BW agents and nuclear materials were available on Voz Island also, in the same way. And much of that WMD procurement will already have been covertly inserted into the U.S. E.g., where do our agencies think that the extraordinarily high grade anthrax mailed in some letters after 9/11 could have come from? Most probably, it came right from Voz Island.

The mailing of high grade anthrax in letters sent to a few important U.S. targets after 9/11 was not an attack at all, nor was it so intended. The mailings were undoubtedly a small *test* attack to ascertain the amount of disruption and economic damage that could in theory be obtained by *mass* mailings simultaneously, through the mail in many locations, to thousands of targets. The results obtained by these little sample mailing tests showed that *mass mailings of anthrax powder, in letters sent to selected high level targets, would create chaos, extensive delay, and substantial economic disruption.* Here again one sees the coordinated planning toward our economic disruption, damage, and collapse.

A Simple but Lethal Strategic Strike with Horrific Casualties

As is known, anthrax is highly lethal when ingested or breathed into the lungs. So a simple spray attack on a large metropolitan area, by a small 2-seat Cessna aircraft with two men aboard — one flying and one spraying out the side — could account for from one to three million casualties {lxxvi}, almost all of whom will die. Quoting Betz {lxxvii,}:

> "A 1993 study by the Office of Technology Assessment concluded that a single airplane delivering 100 kilograms of anthrax spores — a dormant phase of a bacillus that multiplies rapidly in the body, producing toxins and rapid hemorrhaging — by aerosol on a clear, calm night over the Washington, D.C., area could kill between one million and three million people, 300 times as many fatalities as if the plane had delivered sarin gas in amounts ten times larger."

Needless to say, the emergency treatment capabilities of the entire struck area would be completely overwhelmed. In such a case, *triage* becomes the official policy. Unless the treatment were started immediately for a civilian who inhaled the anthrax, it would be largely ineffective anyway. *Once the symptoms appear, conventional treatment is largely hopeless.*

Triage is the classification (medical screening and sorting) of patients to determine the need and priority for treatment and transportation {lxxviii}. Triage generally separates patients into four categories, according to the priority for treatment that will be followed:

High Priority (Red): Patients who need immediate treatment and immediate transport in order to survive. (But refer also to category 4 below). Those who need immediate treatment but cannot be expected to survive even if treated go into category 4 (Black), not Red.

Intermediate Priority (Yellow): Patients who will most likely survive but require treatment.

Low Priority (Green): Patients who require little or no treatment or whose treatment and transportation can be delayed.

Lowest Priority (Black): Patients who cannot be expected to survive even with treatment, those who cannot be expected to survive in a mass casualty situation, and those whose vital signs are absent.

In a mass casualty situation with very scarce supplies and nowhere near sufficient mass treatment capabilities, triage becomes necessary and officially applies. That means casualties classified in the Black (lowest priority) category — such as patients already showing the anthrax symptoms, and therefore almost certain to die with or without treatment — will just be set aside, made as comfortable as possible, and *allowed to die.* There is very little to be gained by treating them with conventional treatment. Even if given every treatment available, almost all of the patients in this category would die anyway. With triage sorting, the very scarce and precious antibiotics and treatments will be reserved for those who have a better chance of surviving if treated.

This is one of the harsh decisions, of the type formerly reserved for military forces in the field under desperate circumstances, which will have to be made for our civilian populace and wounded as well.

The True Cause of Gulf War Disease

By use of scalar interferometry (as is possessed by the Yakuza) or quantum potential weapons (which the Yakuza does not have, but the KGB/FSB does), in advance of the attack it is possible to trick the immune system of each person in the targeted populace area so that it erroneously "detects" invasion by, say, two dozen different pathogens at once. This is done by simply placing the EM "shadow disease engines" or "shadow precursor engines" (in the virtual level just beneath quantum level) for those specific pathogenic conditions in weak EM radiation or in quantum potentials focused into the bodies in the area targeted.

The immune system reacts to what it detects, not necessarily to what is actually occurring. Given this deliberate stimulation and the resulting alarming false detection, the deceived immune system desperately spreads its finite assets across those *"detected" two dozen invading "shadow" pathogens*. And it does detect the interaction of faint precursor engines for the specific two dozen diseases, deliberately radiated into the body and thus interacting with charges in its cellular mass. *Such a "spread" immune system is then just about totally vulnerable to any real attack by any normal opportunistic real pathogen in the area — either natural or terrorist-induced — because the immune system cannot marshal any additional resources.* In a bizarre way, this is a new form of asymmetric biological warfare, designed to pre-disable the immune systems of the targeted populace, dramatically increasing the effectiveness of the real BW pathogen being introduced (by anyone, including other terrorists).

In the absence of deliberate BW pathogen attack, the result of spreading the immune systems in a targeted area is an extraordinarily heightened vulnerability of the radiated populace to ordinary opportunistic infectious pathogens in the populace's environment, to which the humans in that populace are then physically exposed.

There are *always* such opportunistic pathogens in one's environment, but usually the immune system can handle most of them with a shrug. But the *deceived living body with a spread immune system* will be quickly and easily attacked by multiple environmental pathogens, unable to effectively resist, and without the organism's usual immune system's protective resistance {lxxix}. *That living body with its spread immune system will shortly develop a whole "cocktail" mix of opportunistic diseases.* If the immune spreading intensity is lowered a bit, then only a percentage of the persons with the effect will develop the "cocktail mix of diseases" syndrome. If the immune spreading is lowered even further, then only those persons already having additional "immune system lowering" will be affected.

This "spreading of the immune system" is what was deliberately done to generate the "Gulf War Disease" in some of our soldiers in the first Gulf War. The immune systems of our soldiers in certain areas — already weakened a bit by multiple vaccinations — were spread by hostile forces, as a test to see if our scientists and analysts recognized what was being done. The immune spreading signals were deliberately lowered to the point that none of the French troops — who had refused to take the mix of vaccinations we administered our own forces, and which do reduce the ability of the immune system for some weeks — developed the "cocktail mix" diseases and syndrome, while some of the U.S., Canadian, and British forces who did have lowered immune systems as a result of their vaccinations did develop the "cocktail mix of diseases" called the *Gulf War*

Syndrome. The natives in the area had not had their immune systems lowered by vaccinations, and so no change occurred in that populace.

Our scientists did not understand it at all. They did not understand it then, they do not understand it to this day, and they will not understand it tomorrow unless forcibly ordered to meticulously consider and experimentally investigate what happens when immune systems are indeed spread in the fashion being advanced, and when specific tailoring to select the targeted populace subset is also used.

Combining immune system pre-spreading with subsequent real BW attack will increase the yield of the attack itself by a factor of at least three. At the same time, other opportunistic infections will also interact very strongly, with little resistance, so that a "cocktail" mix of diseases will ravage the population whose immune systems have been spread.

We point out but do not further discuss the dramatic effect on those treated Triage casualties, if those casualties have pre-spread immune systems. The effectiveness of the treatment will be severely lowered, so that a much greater percentage of those treated will also die.

Our scientists still have not comprehended the higher group symmetry electromagnetic mechanisms by which health and disease changes in personnel in the U.S. Embassy in Moscow were induced during several decades of microwave radiation by the Soviet Union. They do not intend to further try to understand it, and they have simply buried it. It does not matter that three U.S. Ambassadors eventually died as a result, or that hundreds of persons were made sick or developed real diseases. Also, our scientists do not know how to properly investigate and explain the directly related Kaznacheyev experiments {lxxx}, nor do they intend to. As a consequence, they have already guaranteed that BW attack on the U.S. populace under the conditions stated above will be spectacularly successful, while U.S. treatment of the mass casualties will spectacularly fail.

This appalling situation further buttresses our previous comment that, not only will our scientific community fiddle while Rome prepares to burn, but it is also assuring that it burns completely.

Terrorism by Electromagnetic Biological Warfare

Not comprehending "immune system spreading" or how it is induced, our medical scientists certainly have no knowledge of the electromagnetic use of specific precursor engines for a specific disease — such as anthrax — *to directly induce the disease in targeted bodies in an entire great area.* Instead of the passive measure of immune system spreading, the same EM weapons can directly transport and induce the precursor engines for developing the anthrax conditions in the bodies in a large targeted populace — such as the greater Washington D.C. metropolitan area, which was the area considered in the referenced Government study involving a spray anthrax attack {lxxvi}.

For example, the exact structural pattern of the necessary *disease induction engines* for anthrax can be directly "recorded" from the EM emissions of sickened human cells or humans with anthrax. Such ordinary EM carrier emissions do contain those specific anthrax EM disease engines as infolded Whittaker patterns {xxi, xxii} — particularly in the infrared and ultraviolet spectrum, and particularly when both spectral ranges are considered simultaneously (i.e., when a full harmonic interval of radiation is considered, and especially when that harmonic brackets the visible spectrum and the targeted organism is in the dark to remove possible interference from the intervening visible light spectrum). Our scientific leaders have little or no understanding of the Kaznacheyev experiments, and most of them simply are not interested in understanding them, since it would require thinking in terms of a higher group symmetry electrodynamics, not the common old seriously flawed electrical engineering.

Hence they do not even conceive of the result of electromagnetically enhancing a strike by a BW agent such as anthrax, or of electromagnetic biological warfare, where the disease itself is induced in the body by the exact disease engine that pathogens carry, but mostly without the intervening physical carrier. This "direct induction of the specific disease condition" via tailored induction of the appropriate specific EM disease engines in the bodies of the targeted populace, is another gargantuan force amplifier that can be either used alone or in conjunction with the previous force amplifier of spread immune systems. The U.S. has no defense against it, nor do our medical science leaders and establishments intend to develop such defenses. Terms such as "immune system spreading," "precursor engineering," "precursor disease engine," and "electromagnetic biological warfare" are not even in their lexicon.

As a force amplifier, the combination of the two force amplifiers — immune system spreading **and** EM BW induction of a single disease simultaneously — can increase the overall casualties from an anthrax attack by from 5 to 10 times what would be obtained by the basic anthrax spray attack of physical anthrax spores alone. An influenza epidemic, e.g., can be made horrific beyond all present expectations, as can a smallpox outbreak, etc.

Further, the present physics of the vacuum already prescribes the basis for precursor engines and their use, including their use in induction or augmentation of disease, or in elimination of disease. Quoting Aitchison {lxxxi}

> "Vacuum polarization, in general, alters the effective force law. Forces, in quantum field theory, are understood as being due to the exchange of virtual quanta... In the case of QCD and QFD, ... a crucial new feature is that the force-field quanta themselves carry the 'charge' of the force-field, i.e. it is as if the photon of electromagnetism carried electromagnetic charge."

Notice, however, that even Aitchison failed to recognize there were no force fields in space, but only the precursors, and that precursor engineering is involved. Every chemical and biochemical reaction emits characteristic EM radiation — based on the making and breaking of electron bonds etc. as modeled by ordinary chemistry. Even these radiations could be recorded and played back, inducing disease, without even going into the more complicated explanation of precursor states and virtual disease engines.

The upper limit of the force amplification (more simply, the *gain*) achievable by precursor attack combinations is of course the limit entailed by the size of the targeted population. Medical and emergency personnel in biohazard protection suits and "deeply protected facilities" are just as vulnerable to this mode of attack as the unprotected populace. So devastation of the treatment teams and facilities also results. Further, even decontamination spraying to kill the anthrax spores would not mitigate the damage already done by EM BW.

The combination of force amplifiers, coupled with an anthrax spray attack even much weaker than that in the Congressional study, can effectively infect and sicken almost the entire population, with nearly everyone becoming casualties and dying. *EM BW thus becomes a strategic force amplifier of strategic "total knock-out" strike implications.*

In a mass casualty situation, the presence of these force amplifiers will also move the vast percentage of the stricken casualties to triage category Four (Black). Casualties in the Black category will receive no treatment, but will just be allowed to die — even if some vestiges of the treatment teams and facilities survive. The use of the force amplifiers means that a much greater percentage will be classified Black and be allowed to die without treatment under normal triage rules.

But our own National Institutes of Health (NIH) and medical science community — and our own intelligence community in that area — are simply not interested in such bioenergetics weapons already developed and tested. Nor have they apparently done even elementary homework in the area of higher group symmetry electrodynamics. Most do not even recognize the term! Nor have they funded any specific higher group symmetry electrodynamics research in the area, outside the terribly crippled U(1) electromagnetics and electrical engineering, *even though the disease force amplifier capabilities of higher group symmetry electrodynamics have already been repeatedly demonstrated, first by the KGB against the U.S. Embassy in Moscow in the former Soviet Union and then by the KGB/FSB in Russia against our troops in the first Gulf War.*

The FSB/KGB already knows and plans such force amplifying measures, which will likely be unleashed upon us to dramatically augment any milder conventional BW attacks. *The nature and magnitude of the asymmetric war to be unleashed against us is dramatically different from what our society knows, expects, and plans for.*

Sadly, most of our present scientific medical research is a vast wasteland as far as any study of (1) scalar interferometry use in disease induction or force amplification, (2) the mechanism used in the decades long EM induction of health changes and diseases by the long so-called "microwave radiation" of the former U.S. Embassy in Moscow, (3) the spreading of immune systems of a targeted population, and (4) electromagnetic induction of specific disease in a targeted populace via the induction of the specific precursor engine for that specific disease. Further, our medical scientists' knowledge of higher group symmetry electrodynamics appears to be nil. Worse, the organized medical science community appears to have no concern at all over its own total lack of knowledge in such higher forms of electrodynamics. Possibly, a part of this nonchalance may be due to — or influenced by — the important financial exchanges that have long occurred in the past between some NIH scientists and outside industrial cartels intent on keeping medical science within the bounds presently imposed.

But as a step in the right direction, NIH has imposed stricter rules on its scientists with respect to their consulting to, or employment by, outside drug companies which in turn are receiving payments from NIH funding {lxxxii}. The agency has even proposed banning all its scientists from doing any consulting work with drug companies for a year {lxxxiii}, while the agency's investigations continue. Recently a real clampdown has been ordered, effectively closing this questionable avenue. Let us fervently hope that this cleanup continues, so that NIH really does again function as an independent government agency dedicated to the benefit of the American public, not just for the benefit of the big drug companies. *At this point, every little bit helps!*

Unethical Release of U.S-Made BW Mycoplasma

Sometimes, there also appears to have been collusion between the NIH and rather illegal (and unethical) testing on human citizens without their consent. As an example, our NIH leaders are still doing absolutely nothing at all about the modified mycoplasma BW agent {lxxxiv} that our own BW researchers developed in the 1950s to burrow inside red cells, feed off their hemoglobin shells, and dramatically decrease the oxygen transport capability of the red cells in the blood. This modified BW mycoplasma agent was developed, and then it was thinned by about 2,000 times and sprayed into the populace and into the environment by combined U.S./Canadian teams in the 1950s in Canada. It was also sprayed by U.S. agencies in Florida.

These sprayings caused the deliberately-manufactured BW agent to enter the mosquito population and thence into the human population of North America. Indeed, in the cooperative joint U.S./Canada program, the modified mycoplasma agent was deliberately implanted in mosquitoes by the government researchers, and those mosquitoes were then released, to ascertain the "spread effectiveness" of the mosquitoes as disease vectors. That's rather like firing some machinegun bursts at random into a very large crowd, to determine the average pattern of "hits" and "kills".

The author personally contracted that specific biological warfare disease — courtesy of our own BW scientists and the inane officials directing the tests, who carelessly assumed that dilution rendered the mess safe — while in the U.S. Army, residing in Quebec, with duty station at the Canadian Armament Research

& Development Establishment (CARDE) at Valcartier, Canada in 1966-68. I survived three serious hospitalizations in Canada only by the final Herculean efforts of the French doctors in Quebec, who opened both the chest cavity and the abdominal cavity and moved the internal organs around for about 3 hours, aerating the tissues thoroughly and unknowingly killing the exposed surface mycoplasma. After that oxygenation of the tissues, the mycoplasma already burrowed inside the red blood cells survived and then went dormant, slowly increasing over the next three decades. I survived, but with a time bomb ticking away inside my body.

Left with slowly increasing chronic fatigue syndrome over the next 33-year period, I even managed to serve a tour in Vietnam and still no one knew what was wrong. No medical tests or examinations showed what it was. I retired from active military service at the end of 1975, and was employed in the aerospace industry in Huntsville, Alabama.

Then in spring 2001, resurgence of the mycoplasma generated runaway heart fibrillation and a heart attack. When the terrorist attack of 9/11 struck a stunned America later that year, I was still struggling to stay alive from the heart attack, with runaway heart fibrillation, and significant hypoxia from my own prior encounter with a "terrorist BW attack", where the terrorists were our own fellows!

As fate would have it, a colleague with the equivalent of a Ph.D. in biochemistry who checks on me about once a year then called; he has been in many "deep black" programs, most of which do not officially exist. He knew of the mycoplasma BW agent development program and the mycoplasma spraying in Florida and Canada. He immediately recognized that the strange malady contracted in Canada in 1968 was actually that BW mycoplasma.

He also told me where to get definitive information on the web and from Senate hearings, and voila! The puzzle was solved. Finally in Dec. 2001 a specific lab test was obtained and the chronic BW mycoplasma infection confirmed — after 33 years of puzzlement, doubt, and slowly increasing physical debilitation. The resulting treatment was a year (2002) on antibiotics, which did annihilate the mycoplasma that were hiding in the red blood cells. But it did not and cannot undo the damage already done throughout my body during those 33 years. As my doctor stated, *"You name it, it's damaged."*

After such a long-term case, one has permanent chronic fatigue, cannot walk very far or stand very long at all, and one takes blood thinner for the rest of one's life to try to prevent strokes and blood clots. One also takes medication to try to control the heart fibrillation and stay alive. The eventual expectation is to go on a pacemaker, then eventually have another unexpected heart attack or stroke, finishing the process. I've also now developed pulmonary fibrosis.

The problem is that, if one *takes* all the medication necessary to control the heart fibrillation, it also reduces the volume of blood pumping, which places one back in hypoxia, leading to strokes and/or heart attacks. If one *does not* take the fibrillation control medication, one reverts back into hypoxia from the fibrillation, which leads to strokes and/or heart attacks. *With the usual treatment by well-meaning heart doctors, one is damned if one does and damned if one doesn't.* They do not and will not prescribe remedial oxygen, since they do not consider any possible involvement of mycoplasma in heart problems and attacks. Attempts to get the NIH to notify the heart clinics and doctors of this terrible problem have fallen on deaf ears — probably because of the NIH's original involvement in producing the modified mycoplasma and unleashing it upon an unsuspecting and trusting North American populace. Hence the NIH does not wish to admit that such programs ever even took place with its participation.

The only way out of the dilemma is to take the medicine to defeat the fibrillation, and also take extra oxygen to prevent the resulting hypoxia. So I take remedial oxygen and pay for it on my own volition, since the heart doctors know nothing of the modified BW mycoplasma and they do not wish to hear of it. Nor does one's insurance company, Medicare, the Veteran's Administration, etc. Nor does the NIH. All those fine agencies simply wish that the disabled veteran with such BW-modified mycoplasma would just die and get out of the way.

By taking one's own remedial oxygen, one remains just barely above the hypoxia zone, at least until something else unexpected occurs to make things worse. That way, one remains alive a while longer, perhaps even for a few years. Even so, one has acquired a delayed death sentence just waiting to happen at any moment. And one remains with severe chronic fatigue, weakness, etc.

Perhaps then it is understandable that the present author has a very personal interest in that particular BW mycoplasma disease and that set of incidents, and in the continuing gross failure of the NIH to acquaint heart doctors in the U.S. of that nefarious program and what happened. Today there is a direct involvement of this BW-modified mycoplasma in a significant fraction of patents with heart disease, chronic fatigue syndrome, etc. Many of the citizens of the U.S. and Canada already have that BW organism infecting their bodies, like a ticking time bomb waiting to erupt perhaps many years later. To this day, very few hospitals and medical clinics are capable of doing the test that can detect the very specific type of BW-modified mycoplasma.

Apparently the NIH considers any contact on this problem — or a similar BW problem — a hot potato to be handled by its "policy" section (i.e., "spin control" section). Bluntly, the attitude seems more concerned with political control than with saving the lives of countless American citizens who unknowingly continue to be afflicted with that BW-modified mycoplasma, continue to die of its complications, or are disabled and their lives shortened because of it. *NIH political orientation seems to be one of the major shortcomings of the U.S. in its required defenses against the biological warfare aspects of the "new" asymmetric war. It also appears that the NIH directly participated and cooperated with the U.S. BW organizations that developed and unleashed the BW mycoplasma in the first place. If so, the misdirected attention of the NIH to cover up will probably continue, and that will help guarantee much of the success and increased mass casualties resulting from professional terrorist BW strikes against the U.S. populace in the near future.*

Ironically, that nefarious modified mycoplasma is a strange kind of BW "force amplifier" that the Soviet BW weapons labs initially developed to "soften us up". Eerily, they themselves did not have to introduce it into the North American populace. Instead, we ourselves developed it and — in what has to be one of the most stupid and asinine BW actions of all time — we deliberately unleashed it into our own environment and against our own North American populace. In comic relief, one is tempted to remark that *"With friends like that, who the hell needs enemies!"*

No Effective Treatment for Mass Casualties

Sadly, in this present asymmetric warfare problem, we are forced to conclude that the NIH and the Centers for Disease Control will not do the research necessary to develop adequate response therapies and means for quick, effective treatment of mass BW warfare casualties. Seemingly they continue to prefer Category 4 triage — just letting most of the patients die — to developing unorthodox but real mass treatment capabilities. None of the associated agencies appears likely to develop anything innovative in this area, and *particularly* not a portable *higher group symmetry electromagnetic* treatment unit {lxxxv} that could effectively treat those millions of Category 4 Triage casualties that will result and die in our cities in the near future.

As we mentioned, there is a *precursor EM engine* for every disease and every debilitated condition of the living body. That means there is also a *precursor EM anti-engine* that can be formed, *amplified*, and introduced into the diseased patient — and that will very quickly reverse the disease condition *in situ* and cure the disease. Because the anti-engine is electromagnetic and can therefore be amplified, the treatment period can be reduced to circa five minutes of non-ionizing radiation by the amplified specific disease anti-engines. In such treatment, the body's own potentials and fields thus have the anti-engines kindled inside them, and after the treatment is removed, the body's own "activated" fields and potentials — at every level — now continue diminishing and destroying the disease engine, and reversing the patient's condition back to normal.

Further, anti-engine patterns for a variety of expected and known BW agents can be developed and incorporated into such treatment machines, and the machine can then just be "switched" to the appropriate anti-engine-generating radiation spectrum required to treat the patient's BW condition (destroy the patient's specific disease anti-engine and reverse the patient's condition back to normal). *The radiations from the machine are after all precursor radiations, whether or not the scientists using the machines know* it.

As a further example of U.S. medical orientation, in 1998 the author, with a colleague, tried very hard to persuade the NIH, DoD, USAF, and other federal agencies to develop a portable unit for such EM mass casualty treatment in a crash U.S. government program, to save millions of lives in the coming asymmetric warfare debacle {lxxxvi}. The EM mechanism proposed would have generated an amplified anti-engine for a specific BW-induced disease. "Pumping" the diseased body with opposing longitudinal EM waves will form amplified anti-engines to a specific disease engine in the stricken body. By then radiating the body with the *amplified* anti-engine, or simply pumping the body to produce the amplified anti-engine, it overrides the actual disease engine, eroding away the disease engine and the disease along with it, and reversing the physical condition back to normal. Or, in biological terms, this is the secret of dedifferentiation of a diseased or disordered cell or cellular body back to normal cells and normal health. It is quick, it was thoroughly proven in France in the 1960s and 1970s, it was evidenced in Becker's work, and it was documented in the hard French medical science literature. *Once developed, this would be a mass casualty treatment therapy far superior to anything presently existing, which could save millions of lives in the debacle now roaring down upon us.*

The NIH, DoD, USAF, CDC, and other federal agencies had not the slightest notion of what we were talking about, and we never got out of the "policy" section at NIH, even with the assistance of Congressman Bud Cramer. Not a single scientist in any of those government agencies cared a whit, or even called to discuss the proposal and its novel physical mechanism. Apparently none checked the hard scientific references we cited from the French medical science literature where the basic mechanism had been proven.

None of those agencies had the foggiest notion that the decades-long microwave radiation of the U.S. Embassy in Moscow already demonstrated the use of higher group symmetry electrodynamic "disease engine" signals to directly generate health changes and diseases at will. They had no concept of the use of appropriate amplified "anti-signals" and "disease anti-engines" to reverse and cure illnesses, or that an extended EM can be developed to either *induce* or *cure* specific diseases to order as one desires.

They certainly do not comprehend the higher group symmetry operation of the human cellular regenerative system — which is responsible for the body's healing — and have not gone much further than Becker's brilliant seminal studies of that system {lxxxvii}. Unfortunately, Becker's work was limited by his use of U(1) electrodynamics. Otherwise, Becker would have had the "anti-engine" EM healing process then and there. Even so, he almost deciphered the control operation of the cellular regenerative system {lxxxviii}. His EM healing method for otherwise intractable bone fractures survived and is still utilized in some hospitals from time to time {lxxxix}. It is simple to understand: an electrostatic scalar potential (common voltage) decomposes into a set of bidirectional EM longitudinal precursor bidirectional wave pairs — in other words, a perfect set of "pump waves" for creating specific amplified anti-engines for the specific disease or disorder. So Becker's silly little DC voltage across a horrible bone fracture actually contained and initiated a most important and specific mechanism for producing specific anti-engines and specific dedifferentiation of the diseased cells back to earlier normal condition.

To show what simple potential (voltage) across an otherwise intractable fracture can do, using its component longitudinal EM pump waves, Becker showed that it dramatically affects many of the red cells that arrive in the fracture area, reconverting those affected red cells to a suitable "precursor" cell (more primitive, earlier cell). The "pumped" red cell shucks its hemoglobin and grows a nucleus — this is *dedifferentiation*, and it is generated by the appositive longitudinal Whittaker precursor wave sets — that constitute the voltage — acting as "pump waves" in the nonlinear optical sense. The pumped bone fracture site has thus become a *pumped phase conjugate mirror* site, with the input or "signal wave" being the specific physical disorder's "engine" already in the stricken body. The amplified phase conjugate output is the amplified specific "anti-engine" for that specific disease or disorder condition.

In accord with nonlinear optical pumping theory, the energy of the pumping waves then establishes an amplified "phase conjugate replica" signal set. This amplified signal (anti-engine) will then quickly reverse the entire physical red blood cell back along its past travel through time in its past short life (red cells only live about three months). Hence the red cell dedifferentiates back to an earlier stage. The "delta" condition in the fracture site involves bone cells. So the red cell is *redifferentiated* by the pumping to move forward to the type of cell that makes cartilage. Then it further *redifferentiates* yet again to the type of cell that makes bone, and that cellular material is deposited in the fracture site, healing it.

In simple language, that is precisely how Becker's bone-healing EM treatment works. The process and its proven experimental success already demonstrates the beginning of the required capability to treat and reverse diseases and physical injury conditions by a peculiar variation of nonlinear optical pumping, using the "disease or disorder engine" as the input signal, producing an amplified "anti-engine". This anti-engine forcibly interacts upon any and all aspects of the cells and the physical body to reverse the situation and restore the body to health.

But the NIH was not interested, even though the elements of this process had also been resoundingly proven by the Prioré group in France, in rigorous work funded by the French government.

Eventually the NIH just shipped the entire package over to the DoD (to whom we had already sent the package anyway) because the package mentioned the Gulf War Syndrome and "that was the purview of the DoD". In short, the NIH wasn't interested in what Gulf War Syndrome really was, or in curing it either. So unless it involves drugs or some form of ionizing radiation treatment or some such, our own NIH was not and is not interested in a proven, demonstrated cure for cancer and other dread diseases, accomplished in France in the 1960s and early 70s, with astounding experimental results of thousands of lab animal experiments already rigorously reported in the French medical science literature but resoundingly suppressed today.

The same U.S. agencies lack any understanding of the mechanism that induced the "Gulf War Disease" and the causes of diseases and health changes at the U.S. Embassy in Moscow.

The entire U.S. medical science community is still using an archaic seriously flawed classical electromagnetics of the material luminiferous ether, rather than the more modern higher group symmetry electromagnetics of mass-free spacetime. Consequently it is far more concerned with the immune system's responses which "physically kill the bad guys", or harsh drugs which "chemically kill the bad guys", rather than the cellular control system responses which actually accomplish all healing within the body after the "bad guys" are already dead.

The immune system and the drugs do not actually "heal" anything — not even the immune system's own damaged cells. The fundamental orientation of American medicine is on physically attacking and killing the pathogens which transport the disease engines, rather than on augmenting and amplifying how the damaged body then regenerates its own cells and reverses its own disease engine via use of higher group symmetry electrodynamics. This strongly prejudices and canalizes our medical science development. It also makes us extraordinarily vulnerable — *helpless* is the proper word — to the coming mass casualty attacks in our cities and population centers.

But to uphold the status quo, our medical scientists seem quite willing to accept that, for really substantial WMD attacks, triage Black will apply on a massive scale. So *they will just let most of the casualties die without any treatment at all. The predetermination to lose all those American lives has already been made, and it seems that nothing will change it.*

Indeed, the entire U.S. electrical engineering community itself does not realize that it doesn't even know or apply what electrodynamics in mass-free space *is*. In the face of such appalling scientific ignorance for more than a century, it is little wonder that our nation is so terribly behind in understanding what can be done — and is being done — by the terrorists using the primary (precursor) electrodynamics in free space, before its interaction with matter. In short, our analysts have not comprehended *Soviet energetics* for nearly a half century, and they have no intention of comprehending it now, even though it is being used against us for powerful strategic weapons we cannot counter.

Our national agencies adamantly are not going to develop the proven Prioré-type EM treatment or really get to the bottom of Becker's anomalous EM bone fracture healing, because such would not be common old electrical engineering. It would also greatly disturb the pharmaceutical cartels, which have bought American medical science by financing research, medical scholarships, medical university expenses, etc. and by providing "consulting fees" etc. to a variety of scientists in the system. Even though most of those cartel companies are benign or even intentionally beneficial, any threat to their dominant positions evokes a sharp response from their evil 1% entourages. One can easily meet with a "sudden suicide on the way to the supermarket" and such.

Our scientific leaders are not going to demand the correction of that terribly flawed electrical engineering model, which is our single greatest national vulnerability and the one that is being capitalized upon by coordinated terrorism and asymmetric warfare to destroy us in within the three year Operations Phase we entered at about the beginning of 2005. We are going to *unnecessarily* lose millions of American lives, with most casualties receiving no treatment at all.

Those future wounded Americans would not have to die if such EM countermeasures were intensely developed in portable treatment machines and massively deployed {xc}. We are describing a vast force amplifier that could be developed for the defending nation that has been targeted and very seriously struck. Our own government and scientific community adamantly will not permit its study or development, and will simply "fight to the death of every one of us" to preclude such a desperately needed therapeutic method and capability,[9] despite rigorous scientific proof by the Prioré group in the 1960s and early 1970s {xci}, as aforesaid.

Impact of Immune System Spreading

Weapons such as scalar interferometers in the hands of the Yakuza have given the terrorists a terrible *force amplifier* of WMD bioweapons that parallels the destruction capabilities of all a modern state's ICBMs, strategic bombers, nuclear subs, aircraft carriers, etc. All without a single ICBM, strategic bomber, nuclear sub, or aircraft carrier.

So for the strategic asymmetric war threat, now one considers — say — three large U.S. city areas hit just about simultaneously by effective anthrax spray attacks, and augmented by EM BW means. One immediately sees the grim picture very probably in store for us. To be conservative, we assume 1 million casualties in each city as the "baseline" achieved by the spray attack pathogen itself. Thus a total of 3 million casualties are obtained in the three strikes, with no other augmentation. However, if immune system spreading were also incorporated simultaneously, the casualties in each city will reach, say, 3 million, almost all of whom will die. That is a total of 9 million American casualties in the three cities — and almost all of those wounded will perish.

---

[9] The leopard has not changed its spots. See Betsy McKay, "CDC Aims to Create Experimental Pandemic Bird-Flu Virus," Wall Street Journal, Wed. Mar. 28, 2005, p. B1, B2. Incredibly, CDC as part of the NIH is still engaging in the deliberate production of potentially highly contagious pathogens, capable of causing rampaging pandemics if unleashed or accidentally escaping. All sorts of "reasons" are advanced. A little analysis from the viewpoint of precisely what FSB/KGB bioenergetics can do in such situations would be enlightening — but then neither CDC nor NIH knows anything about bioenergetics, even after four decades of its testing against us to see if we recognize it. Again, with friends like that, who needs enemies!

And that's three small 2-place aircraft, six terrorists, three little agricultural hand sprayers, 300 kilograms of anthrax spores from Voz Island, and a little gas and time.

If, in addition to the immune system spreading, some supporting scalar interferometers are also inducing the same anthrax disease pattern via EM BW, that's now perhaps 5 million casualties in each city area, for a total of 15 million American casualties, almost all of whom will die. If the intensity of the scalar interferometry disease induction is further increased, that's perhaps 18 million or so total American casualties in those three city areas — almost all of whom will die.

Here one directly sees the transition from the kind of estimates our own analysts usually make for BW agent attacks, to estimates that show the more maximum risk of massive strategic destruction and even strategic knockout of the U.S. by the same "front line terrorist troops", due to the advent of the supporting scalar interferometers, immune spreading, and EM BW.

All that would be caused *overtly* by 6 men, three light 2-place Cessna-type aircraft, 3 little agricultural sprayers which can be purchased at any Home Depot or similar store, and 300 kilograms of anthrax spores from Voz Island — and, in the 9, 15, and 18 million casualty cases — a little force amplification support from three or more scalar interferometers.

But that's not the end of it.

In addition, the contamination of those large built-up population areas will require the ruthless spraying of decontamination agents, killing many thousands more of our own populace. *The effectiveness of the decontamination sprays would also be seriously hampered if the scalar interferometers continue their EM BW anthrax induction. None of these kinds of "force amplifier" effects are presently known or taken into account by our analysts and our intelligence agencies.* They are designing and planning for the type of warfare they themselves know and expect. They are not designing and planning for the type of warfare that is coming down upon us.

The most urgent "action fact" that should be concentrated upon by our medical science agencies is that most of those casualties are going to wind up as triage Black category 4 patients. Most are not even going to be treated, and so almost all of them will die. *The jobs of the leaders of our national health research agencies should be made directly dependent upon them rapidly examining and finding new and revolutionary treatment capabilities that will negate the present "triage 4" massive catastrophe. If they can't or won't do it, replace them!*

And *there are viable candidates to examine.* We have pointed out that the "amplified disease anti-engine" treatment approach is capable of rapidly causing the reversal of the disease, even in those who presently will be assigned to triage Black category 4. Most of those terrible casualties now falling into triage Black category 4 would fall within triage Red category 1 instead, if small portable treatment devices are available by the thousands. There is simply nothing else out there that will produce such a dramatic savings of human lives, and probably the survival of our nation itself.

Impact of Smallpox Release

It is almost inevitable that the terrorists also obtained smallpox BW agents — so could anyone who wanted it, had guts and had money. Reliable information already confirms that the terrorists obtained many BW agents, including smallpox, by going into the BW labs in former Soviet republics, and buying it. Or one got it from North Korea. Or one just hired very desperate Russian smallpox BW scientists who urgently needed to support their families. Or one got some of it from Voz Island.

The collapsing Soviet state was unable to pay salaries to scientists, generals or soldiers for protracted periods. In desperation to get money to survive, those individuals were forced into extreme measures. *Anything there that one wished to buy could be bought, including even a nuclear submarine* — see Lunev's book for an incident {ix} where a Russian nuclear submarine was actually purchased. Among other things, one could obtain smallpox agent and the services of Soviet BW smallpox specialists themselves. And of course it included nuclear weapons.

Once smallpox is unleashed in any major city on the planet, it will eventually kill almost one-third the present human population. Quoting Garrett {xcii}:

> "If the smallpox virus were released today, the majority of the world's population would be defenseless, and given the virus' 30 percent kill rate, nearly two billion people would die."

Note that Garrett also knew nothing about immune system spreading, or direct "smallpox engine" induction electromagnetically, to augment and increase the kill rate. Smallpox will almost certainly be released against the U.S. populace, probably with EM augmentation. The Yakuza and other terrorist organizations will see to it, *with smallpox toxin supplies almost certainly already inserted in the U.S. and waiting*, and with the scalar interferometers already waiting to induce immune system spreading and direct EM BW engine insertion in the targeted populace.

As an intellectual exercise, we leave it to the reader to ponder the impact of a few limited smallpox releases in our cities combined with simultaneous wide usage of the force amplifiers (immune system spreading and EM BW smallpox induction) previously discussed, particularly in the grim absence of any effective mass casualty treatment capabilities. The impact is not pleasant.

The Grim Nuclear Problem

One can also see the damage that can and will be done by dirty weapons (ordinary high explosives surrounded with nuclear materials), spreading nuclear fallout in densely populated areas..

But there is another, even grimmer, side of the story. When the Soviet Union collapsed economically, many nuclear weapons were made available on the black market, and sold to the highest bidder. Some 200 or more nuclear warheads for missiles came up "missing" from the Ukraine alone. Quoting Communist Party leader Petro Symonenko {xciii}:

> "Out of 2,400 nuclear warheads which were on Ukrainian territory, the withdrawal of only 2,200 warheads has been verified. The fate of the remaining 200 warheads is unknown."

Further, in 1996 Alexander Lebed, then Boris Yeltsin's national security adviser, reported that some 100 suitcase-sized Russian nuclear bombs were missing. In a donnybrook, some Russian officials denied the existence of such weapons, while others confirmed their existence. The charges and counter charges faded away, and nothing was ever resolved.

Nuclear weapons do not just "go missing" with a shrug of one's shoulders!

All those missing weapons were in fact sold on the nuclear black market. Guess who bought them, and guess where many of them have been inserted already, waiting to go, along with teams trained to detonate them on command. It is almost certain that Bin Laden and Al Qaida succeeded in purchasing some of the stray Russian nuclear weapons. Other terrorists such as the Yakuza and the Aum Shinrikyo almost certainly purchased some of them also.

Terrorist efforts to obtain further bomb grade material such as highly enriched uranium (HEU) have continued. Quoting Bunn and Wier {xciv}:

> "The 41 heavily armed, suicidal terrorists who seized hundreds of hostages at a Moscow theater in 2002 reportedly considered seizing the Kurchatov Institute instead — a site with enough highly enriched uranium for dozens of nuclear weapons… Al-Qaida has been actively seeking nuclear material for a bomb and has strong connections to Chechen terrorist groups."

Bunn and Wier {xcv} also point out that:

> "More than 130 research reactors in dozens of countries still operate with HEU fuel, and many have no more security than a night watchman and a chain-link fence. Pakistan's heavily guarded nuclear stockpiles face huge threats, from both insiders and outsiders, including large remnants of al-Qaida and the Taliban in the country." … Comprehensive U.S.-funded security upgrades have been completed for only 22 percent of Russia's potential nuclear bomb material; upgrades for tens of thousands of bombs' worth of material are still incomplete."

At this moment, no one knows how many Russian nuclear weapons have gone "missing" and were sold on the world black market by unpaid Russian generals and soldiers desperate for money to feed their families. No one knows how many Russian nuclear weapons are still "going missing" and being sold on the nuclear black market. Because the terrorists wanted them and had the funds, and because the Russian generals and soldiers sold them to someone, it does not take a rocket scientist to understand that the two parties almost certainly did business with one another. For more specific information, see Williams {lxxiii, xcvi}.

The Al Qaida has quietly confirmed to the Arab world that it, the Al Qaida, already has inserted seven nuclear weapons in major U.S. cities, and some of the weapons were 200 KT. At least one U.S. author has assessed the Al Qaida as having nuclear weapons in the U.S. {xcvi}. The present author personally suspects the actual number inserted by the Al Qaida and other terrorist organizations combined is about double that figure, at least. Bin Laden apparently bought some 20 suitcase nuclear weapons himself {xcvii}, using funding obtained from the Arab world and/or drug production in Afghanistan.

Recall that the KGB/FSB faction which controls the Russian superweapons — and the leasing and transfer of scalar interferometry technology to the Yakuza/Aum Shinrikyo — also controls the previously inserted Russian nuclear weapons in American cities, as well as the associated Spetznaz teams. During this orchestrated two year period we have now entered, somewhere in the United States at least some of those weapons are very probably going to be transferred or sold to the Yakuza/Aum Shinrikyo, and perhaps a few will go to other well-organized groups such as Al Qaida.

The capability exists. It serves the purpose of the FSB/KGB in assuring the destruction of America. It will be done.

As another example, the former Soviet Union also probably inserted anthrax and smallpox agents into the U.S., along with teams to unleash them on command. If so, those assets are also still here, waiting for the order to move. It seems highly probable that at least some of these BW weapons caches will also be released to the terrorist organizations such as Al Qaida, for use in the coordinated coming attacks.

So in addition to the BW threat, somewhere between 14 and 40 already-inserted hostile nuclear weapons probably form the "minimum nuclear weapons pool" available to the coordinated terrorists during this two year period. At least that would appear to be a reasonable estimate. We accent that this estimated nuclear weapons pool has — with high probability — already been inserted on site in the U.S. *This is another startling capability of the present terrorists to deliver a resounding strategic nuclear knock-out blow against the United States, without ever firing a missile or launching a strategic bomber.*

These are examples of the true state of affairs concerning the "War on Terrorism". The great battle of that war has already slowly started, with the zeroing in of strategic Yakuza scalar interferometers on the Yellowstone supervolcano (giving them a "single shot knockout capability against the U.S.), and with the subsequent Yakuza induction of the 9.3 seafloor earthquake on Dec. 26, 2004 that induced the catastrophic tsunamis in Asia. Presently additional U.S. volcanoes are active and rumbling, as are hundreds of seafloor volcanoes. And so the action is slowly beginning. The FSB/KGB will gradually unleash these terrorist assets, and then their initially sporadic attacks will be deliberately increased, with maximum effort at the very end of the second year and over the first half of that following third year. The intent is to generate a terrible and catastrophic economic collapse of the U.S. two years hence, and then have the Yakuza methodically destroy our cities and every living American.

Harshness of the Terrible Decontamination Problem

No one has a good solution to decontaminating mass metropolitan areas that have been destroyed or heavily damaged and that are dramatically contaminated with nuclear fallout. Neither does anyone have a solution for decontaminating the *distant* large nuclear fallout areas surrounding those devastated cities after the nuclear detonations.

Consider the similar terrible task of decontaminating a large city (such as Washington D.C. in that government study), once the city has been sprayed en masse with anthrax by the terrorists. Yes, there are some chemical sprays that will kill the anthrax spores (else they last for a century or more, apparently). But these sprays are very harsh. To spray a civilian populace and area with them is to unavoidably kill a certain number of thousands of one's own sprayed civilian populace. Multiple sprayings of each contaminated area will be necessary, in order to attain the required assurance of sufficient decontamination.

So far as is known, there are no similar sprays available that can remove contaminating nuclear radiation!

Now note the extreme severity of the coming decisions necessary in asymmetric war. In war, high level military commanders have always been faced with agonizing decisions. For example, suppose an Army commander has — say — a dozen divisions under his control. Suppose his war front is attacked by overwhelmingly superior forces, and breakthrough and destruction of his Army is imminent. The commander will have to deliberately sacrifice — say — two divisions (that's 36,000 men) by hurling them into the teeth of the attacking forces, to temporarily halt and delay the attack momentum while those two divisions fight and die. They will be ground up just like hamburger meat, but that delays the enemy onslaught for a precious short time.

That desperately-bought short delay time gives the Army commander the chance to get his other 10 divisions out of there, and back to prepared defensive positions, to live to fight another day. This way, he saves the bulk of his army, by deliberately sacrificing those two divisions. *The point is, military commanders have always had to agonize and sometimes deliberately send large numbers of men under their control to their certain deaths. The soldiers sent on such missions also knew when they received their orders that they went to their certain deaths.*

Any fool can make a military decision when there is a "good" option and a "bad" option. But when there are only two "bad" options, one is forced to decide for the "least worst" of the options — even though it entails the deliberate sacrifice of friendly lives to save others.

Now the battlefield is our own cities, and the targeted force is our own civilian populace. That same type of "least bad" decision necessity falls now on the sacrifice of some thousands of our own civilians. Some commander at high level (probably the President himself) will have to assess the situation as so desperate that multiple decontamination sprayings of the anthrax-contaminated area and populace are absolutely necessary, to prevent spreading mass destruction by anthrax carried to other areas, and resulting in the deaths of many millions more. So such an order will have to be given, agonizingly, to spray multiple times (necessary to neutralize the anthrax with a good probability of success).

Each massive spraying might kill perhaps an additional 5,000 or even 10,000 U.S. civilians. Heartbreakingly, these will be the most defenseless and helpless of our citizens, and the ones we would strongly wish to defend to the death with our very lives and with every heartbeat in our breasts. The most vulnerable group will largely consist of persons with hampered breathing or vulnerable immune systems — the old, the feeble, babies, the sick, nursing home occupants, etc.

It has been said that war is hell, and indeed it is. Asymmetric war resoundingly brings that hell home to our civilian populace.

In the past, our troops have mostly ridden out of the protected U.S. fortress/castle to a foreign land, and fought the war "over there", with our civilian populace largely immune and shielded from the horrors of such fighting, particularly when the fighting is intense. Now in addition to our involvement in foreign places, in asymmetrical warfare the major battleground is right here in our own cities, because the major target being attacked by the hostile forces is our civilian populace — including our aged, the infirm, women and children, babies, etc. *These devils who are our foes believe that, if they can kill lots of American civilians, babies, etc., they personally acquire a very high religious status from God Almighty*. In short, we are up against a fanatical, irrational foe, one who has also unfortunately had two decades or more of almost unrestricted access into the United States, to insert the necessary WMD assets for this "final great Jihad" against us in our cities and homeland.

This "final great Jihad", in the operations phase fought inside our cities and nation, is what is really coming upon us, together with a similar grim dedication to our total destruction by the Yakuza. This is the coming horrible nature of the struggle. We are actually in a clash to the death between civilizations and cultures. Quoting Seib[10]:

---

[10] Gerald F. Seib, "Deeper Threat is Being Missed amid Iraq Debate," Wall Street Journal, Wed. 8 Sep. 2004, p. A4.

> "The far bigger question is whether the U.S. is engaged in a historic clash of civilizations with the Islamic world. ...The U.S. ... has reached no consensus on a strategy for dealing with a clash of civilizations."

Seib further states[11]:

> "The bin Laden goal isn't simply to humiliate the U.S. Nor is it to overthrow the government of Saudi Arabia, or Egypt, or Jordan. The goal is to eliminate those governments — in fact, to eliminate those nations."

According to the present Yakuza/KGB/FSB and international terrorist plans, their actions are expected to reduce the U.S. to lying prostrate and helpless two years from now. With a U.S. nation in utter chaos and ruin, and economically already collapsed catastrophically, a touch-up with the FSB/KGB quantum potential weapons (the most powerful weapons on earth, but possessed by at least five nations) would quickly dud every nuclear weapon, ICBM, nuclear bomber, nuclear power plant, nuclear submarine propulsion system, etc. on the planet {xcviii}. The dudding could be achieved in about 10 minutes, although an hour or so seems to be allotted for it. This is "pulling the dragon's teeth and claws." *The conventional strategic military power (ICBMs, nuclear bombs and warheads, and nuclear subs etc.) will be suddenly disabled and effectively wiped out, to exist no more. All that giant strategic apparatus will be just so much trash on the garbage dump of history.*

Further, the U.S. society will be enmeshed in total chaos from the intensity developed in its polarizations. The society will be massively spilling its own blood, fighting in the streets, as a result of conversions of the leaders of the polarization areas. In addition, massive "sleeper" conversions suddenly triggered en masse throughout our government leadership, scientific leadership, police, armed forces, large technical corporations etc., will generate a society instantly gone totally berserk. We shall see police attacking cities and civilians, armed forces attacking each other and the civilian populace, etc. The *order* in all organizations that is necessary for civilization to survive will suddenly have dissipated, never to return again. This is one way in which, by themselves, psychoenergetics weapons can totally destroy an entire culture or the entire Western civilization.

At that point, with a U.S. in total chaos, the Yakuza gleefully plan to participate in the methodical destruction of the United States — the terminal part of the present Operations Phase. It will be a turkey shoot. The Yakuza will be freed to just blast away at will with their scalar interferometers, as they wish, to continue to the full destruction of the United States by destroying cities, infrastructure, and the entire civilian populace.

The intention of this unholy war against us, by both the recognized terrorists and the Yakuza/Aum Shinrikyo is to kill every American (and every Israeli also) —

---

[11] Seib, *ibid.*

men, women, children, babies, all of us. It is actually a war intended to totally destroy our civilization — and every living American.

We will see the operational unleashing of this terrible plan, already just beginning during the two years starting at the beginning of 2005, and gradually increasing in intensity, and finishing at the end of the second year with the catastrophic economic collapse of the United States. Once the economic collapse is complete and chaos is supreme, the sudden dudding of all our nuclear weapons, power plants, and nuclear propulsion systems will be the coup de grace that leaves us a hapless target. *The rest of it is the turkey shoot, to simply execute us and finish us off.*

In the interim, we will see damaging volcano activity increasing, and very probably we will see powerful tsunamis striking both the U.S. West Coast and East Coast.

If it were to somehow become necessary, the Yakuza will also trigger the Yellowstone supervolcano into violent eruption, destroying the United States and most of North America.

Summary and Conclusion

In closing, we reiterate some massive strategic capabilities already registered and waiting. Toward the end of 2004, the Yakuza scalar interferometers registered (zeroed in) on the Yellowstone Caldera, the largest supervolcano on this planet. With that "registration" completed, they now can initiate that supervolcano into violent eruption, whenever they wish. Again, *the supervolcano's eruption will likely expel more ash, lava, debris, and rocks than the entire Grand Canyon can hold, devastating a large part of North America.*

The caldera is overdue for another eruption anyway, so undoubtedly there is increased geophysical stress there already. One should also check the latest findings in geology, and see what happened to North America the last time that Yellowstone Caldera erupted violently. The last eruption devastated quite a portion of North America, killing most of the higher forms of life in a large area.

Now check what happened the last time the somewhat smaller Toba supervolcano in Sumatra, Indonesia erupted violently, about 74,000 years ago. *It destroyed all human life on earth, except for a few thousand humans in Africa* {lix}. The human genetics shows this sudden "bottleneck" in our genetic variability that sudden occurred about 74,000 years ago {lix}. Also, this is why the present human genetics also shows that "we all came from Africa". And so we did; *we are all descendants of that small group left alive in Africa about 74,000 years ago.*

In registering on the Yellowstone caldera, the Yakuza has already achieved an additional tremendous "knockout" strategic strike capability against the United States. The Yakuza can indeed wipe out much of North America in a single strike, then finishing the destruction of the United States at its pleasure. There are of course other volcanoes in the U.S. that can also be stimulated into eruption to add to the chaos and carnage. Several of them are already being "actively tickled".

Again, in 1997 U.S. Secretary of Defense Cohen already confirmed *the use by terrorists of electromagnetic weapons for precisely such purposes* — initiating volcanoes, generating earthquakes, and controlling our weather and climate. The recent September 2004 "shooting" of one powerful hurricane after the other into the United States, is just one example. The economic damage done by those hurricanes Charlie, Frances, Ivan, etc. has been adequately covered on the national news, so it is already familiar to the reader.

Other powerful capabilities available to the Yakuza scalar interferometers are also possible and therefore registered, such as deliberately generating tsunamis that then strike targeted coastal areas at jet liner speed {xcix}. One such "tsunami looking for just a touch to complete its preparations to happen" is provided by the Cumbre Vieja volcano in the Canary Islands. A large section of the volcano has already broken loose, and threatens to slide into the sea. If so, it will generate an enormous tidal wave (tsunami) that will strike the East Coast of the United States a few hours later {lx}.

For example, quoting Bill McGuire, director of the Benfield Grieg Hazard Research Centre at University College in London {c}:

> "A 12-mile chunk of the Cumbre Vieja volcano in the Canary Islands rattled loose during a previous eruption and is at risk of smashing into the Atlantic Ocean, triggering what could be one of the largest tidal waves in recorded history. … It's not a matter of if, but when."

Obviously, if this "submarine slump-induced tsunami" is induced sharply by Yakuza scalar interferometry, so that the slump occurs with sufficient force, then 8 to 10 hours later a great tidal wave 70 feet or more in height (conservative estimate! Some estimates run 200 feet high) could strike New York and the Eastern U.S. Coast, as well as the Caribbean, moving at some 600 miles per hour. Such a giant tsunami would penetrate inland and devastate entire coastal areas. *It is an opportunity that will not have gone unnoticed by the Yakuza, and undoubtedly one or more scalar interferometers have already registered on that target.* Again, we strongly stress Defense Secretary Cohen's confirmation of just such use of electromagnetic weapons by terrorists. This is one more example of what can be done by "deliberately evoking, assisting, and directing already available giant geological forces".

So weather engineering and steering and augmenting hurricanes and other storms is just one facet of the real threat. But it is a very important part of the *strategic* threat, as indicated by Secretary of Defense Cohen in 1997.

Hostile Yakuza fingers are already on the triggers of those zeroed-in scalar EM interferometry weapons, 24/7, waiting for the order to be given to strike. If deemed necessary, this already includes the eruption of the Yellowstone supervolcano and the massive tsunami hitting the Eastern U.S. seaboard from induced violent eruption of the Cumbre Vieja volcano.

These two examples are the kind of things that, to one extent or another, are now coming down the pike at us full-speed.

We have also briefly pointed out the horrific casualties that can be obtained by very small and simple anthrax attacks, when bolstered by force amplifiers (immune system spreading and EM BW disease induction).

As can be seen, once the superweapons (such as scalar interferometers) already in the hands of the asymmetric warfare terrorists (the Yakuza component) are factored into the analysis, indeed the terrorists have already achieved a strange kind of "rolling thunder" strategic knockout capability against the United States. At least publicly, this major change in the capabilities of the terrorists against the United States has not been analyzed or pursued by the conventional U.S. government agencies and their traditional support analysis agencies. Specifically, the U.S. Congress has not been adequately informed, particularly of the terrible mess that our decrepit old electrical engineering really is, and the consequences of not correcting it when our foes already have done so.

Finally, the reader is strongly urged to review in detail the section on psychoenergetics weapons and scenarios, in the second part of this book. Massive conversions will be used against the U.S. and Western nations, and it alone can in fact wipe out Western civilization. Hundreds of scenarios are available; several examples have been chosen and will be discussed. Massive insertions of "sleeper" personas in massive numbers of individuals in the targeted U.S., Israel, and Western Europe are already underway. The nonmaterial precursor robots, which were launched against us in an eerie and officially undetected giant war against us at the end of 1999 and first few days of 2000, are once again coming at us en masse. Again the reader is strongly urged to check the information on that potent threat in the second part of this book.

When the strategic strikes and new kinds of force amplifiers are factored into the war with terrorism, we find that we now confront a den of raging but carefully controlled tigers, not just a den of small cats using car bombs and solitary humans with strapped on C-4 explosives. Instead, we confront a coordinated strategic threat of awesome and unprecedented power and tremendous damage capability, one that already has the capability to destroy us several times over, and one that is slowly tightening the noose about our collective necks in a deliberate three year schedule ending in our total destruction and oblivion.

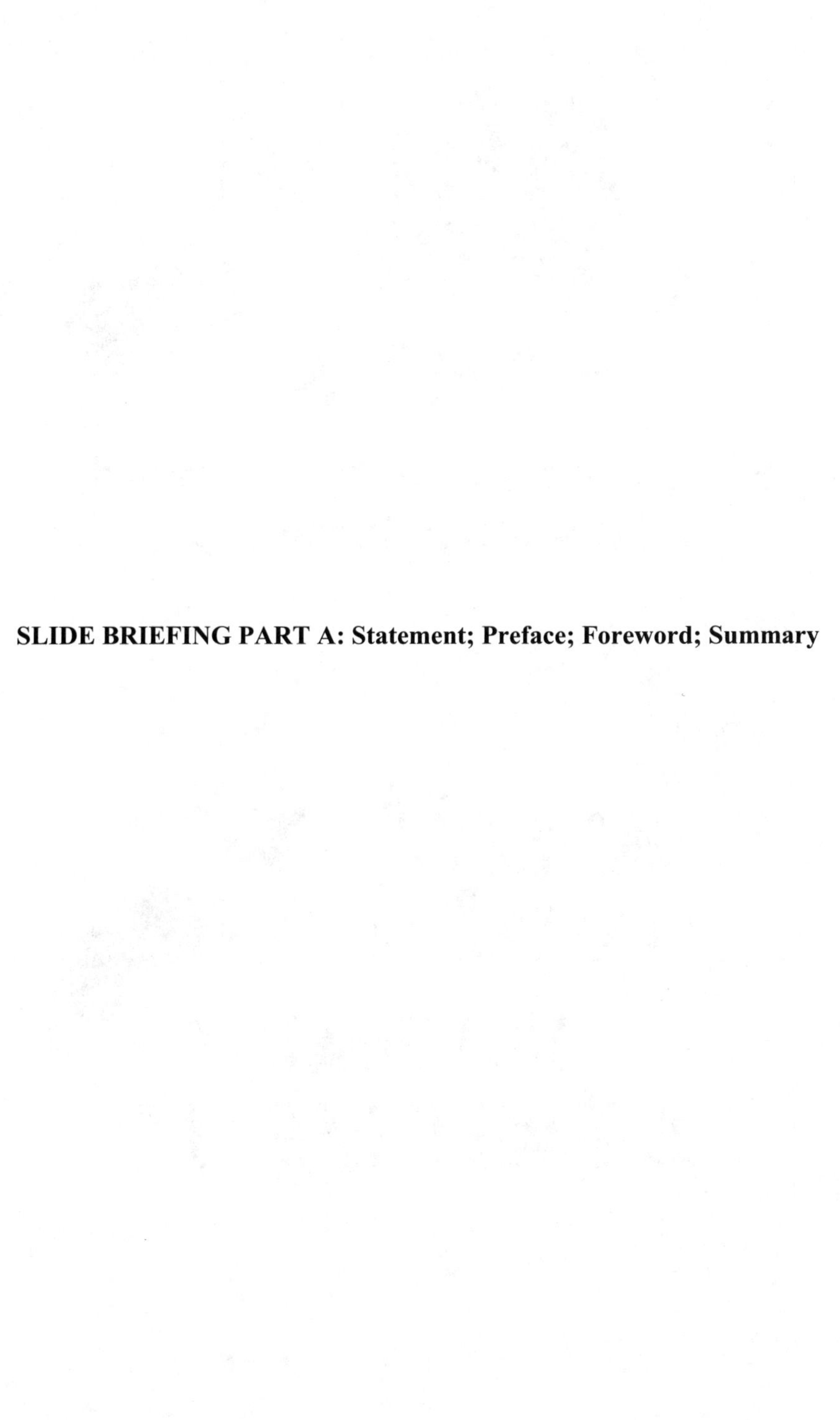

SLIDE BRIEFING PART A: Statement; Preface; Foreword; Summary

PART A

Statement, Preface, Foreword, Executive Summary

INITIAL STATEMENT

Statement of the situation

- Since the late 1970s, Israel has repeatedly saved the United States from pending destruction.
  - 1970s, saved us from already-inserted Spetznaz teams with nuclear weapons, who would have blown up our cities.
  - 1986, saved us from pending Soviet attack with energetics weapons, particularly scalar interferometers.
  - 1997, saved us from Soviet attack twice, which would have been attacks by the full energetics weapons arsenal.
  - 1999-2000, destroyed millions of nonmaterial robots in subspace inside EM signals, fields, and potentials throughout the U.S.
- Presently, attack is again scheduled, with addition of the eerie new mind conversion techniques of mind war.
- Israel *may* be able to save us once again, in which case we shall survive. Otherwise, we will be destroyed, beginning about two years from now.

Mind Wars: Catastrophic end of civilization?

- *"I firmly believe that before many centuries more, science will be the master of man. The engines he will have invented will be beyond his strength to control. Some day science shall have the existence of mankind in its power, and the human race [shall] commit suicide by blowing up the world."*

  Henry Adams, 1862

- COMMENT: With perfection and unleashing of psychoenergetics weapons, we have reached that time envisioned by Adams, in little more than a century.

End of technological species: The final catastrophe (1)

- Some decades ago, we pointed out that a technological (tool-using) intelligent species eventually poses its own self-destruction, by the ever-increasing power of its tools and by its inbuilt inhumanity of one human member for another.*
- Either the technological species uses its eventual technology to *transcend* the inhumanity of one human for another, or it uses it to *utterly destroy itself* because of its innate inhumanity.
- It is not accidental that we find no signs of intelligent life on other distant planets and in distant galaxies.

End of technological species: The final catastrophe (2)

- Today the human technological species has developed and tested eerie superweapons that breach the threshold required to move into the final decision domain. We have entered that domain already.
- Accordingly, within the next three years we face coming calamities so appalling they numb the mind and nearly defy the imagination.
- Our most basic human element — the ability to function mentally as a unitary individual or group — may disappear forever, destroying the species.

---

* T. E. Bearden, The Excalibur Briefing, Strawberry Hill Press, San Francisco, 1980. Updated edition, Cheniere Press, 2001.

GO: Group coherence and Order (1)

- The functioning of the human species — from the individual to large nations to the entire species — is determined by the ability to *group coherently* and thus by the ability to *order (cohere)*
    - Restricted loss of GO in individual is *neurosis* (sickness).
    - More general loss of GO is *psychosis* (insanity).
- Neurosis and psychosis also exist in the group.
    - Restricted loss of GO in the group is *group neurosis* (sickness).
    - More significant loss of GO in the group is *group psychosis* (insanity).
- Complete loss of GO in the individual or group is total chaos and total insanity (the old notion of hell).

GO: Group coherence and Order (2)

- FSB/KGB developments in *psychoenergetics* provide the ability to affect, reduce, and eliminate GO in the individual, a small group, a large group, a nation, or a group of nations.
- The technology provides for reducing or eliminating GO at any or all elements of the attacked society. In effect, for complete "dissociation" of the society.
- It is being unleashed, eventually massively, spelling the rapid demise of the U.S. and Western Europe.
- The genie cannot be rebottled. Ultimately it may destroy GO for all human society. If so, the dissociated human species rapidly consumes itself and dies.

Development and destruction of the human species

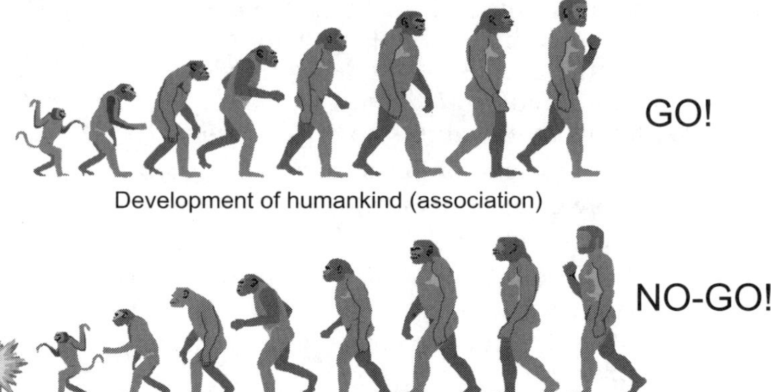

Development of humankind (association)

Destruction of humankind (dissociation)

End of technological species: One possible catastrophe (1)

- Some 74,000 years ago, a great supervolcano at Lake Toba on the Island of Sumatra violently erupted.
  - Destroyed all humans on earth except for a few thousand in Africa.
  - *All present humans have descended from them*.

LAKE TOBA

- Human genetics records the severe "genetic bottleneck" in human genetic variability that occurred.
- Hence present human genetics shows that "we all came from Africa".

End of technological species: One possible catastrophe (2)

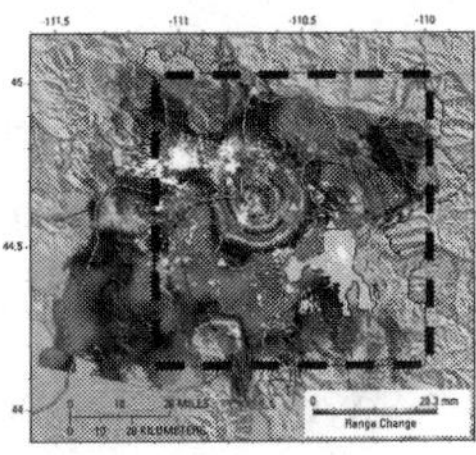

YELLOWSTONE CALDERAS

- Underneath Yellowstone National Park in the U.S. lies another great supervolcano, with 3 calderas, and it is overdue to erupt.
- In 1997 SecDef Cohen *confirmed* terrorists' use of "EM weapons" to cause earthquakes, erupt volcanoes, and control our climate and weather.
- The Yakuza zeroed-in large scalar interferometers on Yellowstone in 2004; can evoke its eruption at will, destroying the United States in a single shot.
- The Yakuza generated the 9.3 undersea test quake in Asia on Dec. 26, 2004 which caused the catastrophic tsunamis in Asia, killing more than 300,000.

End of technological species: The final catastrophe (3)

- With the rise of energetics, FSB/KGB superweapons have been developed which target material systems, living bodies, and living minds in ways never before thought possible.
- To create, engineer, shape, control, or destroy living human minds at will — including all of them on Earth in one strike — uproots the basis for all human order.
- By destroying coherence of mind and minds, no human or organization can survive or function.
- It poses unthinkable catastrophes exceeding any past disaster the human species has experienced.

End of technological species: The final catastrophe (4)

- These psychoenergetics weapons have been developed, tested, and deployed en masse.
- We have entered the last three years of the gradually escalating Operations Phase of asymmetric war against us, which will little resemble anything we have known and experienced before.
- We may see catastrophic destruction of civilization itself, and perhaps of all human life on earth.
- Our purpose is to point out the elements of the coming catastrophe, in the hope that somehow it can still be avoided or at least lessened in severity.
- Otherwise, the coming debacle may soon finish what Toba almost accomplished 74,000 years ago.

Preface

Some major concepts

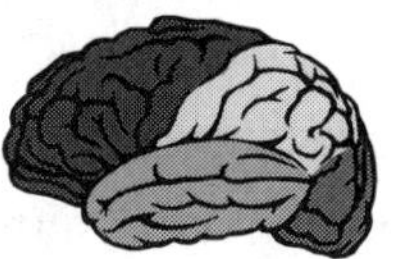

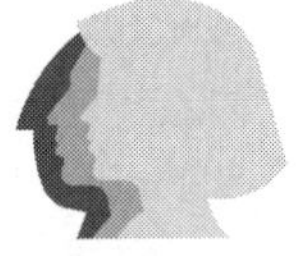

Mind war is the use of *conversions*

- The Creator created life and living biological systems.
  - Created the *process for making* life and living systems.
  - *Science can discover and engineer this process.*
- FSB/KGB scientists have done precisely that.
  - From a distance, can insert altered variant of targeted mind in targeted individual's body.
  - Can switch between the two ("conversion").
  - Changes the behavior designed for targeted individual, or any number, as desired. Conversions against U.S. and others (such as Israel) underway.
- Conversions, fractional conversions will play a major role in U.S. economic collapse two years from now, and in the destruction of the U.S. that will follow.

A living mind can be produced and formatted, as can a variant of it (1)

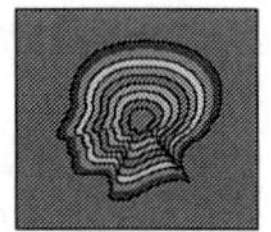

- Make three "precursor engines" (sets of ST curvatures).
  - One for mind dynamics on time axis.
  - One for body dynamics in 3-space matter.
  - One for the coupling dynamics (4-spatial).

- The three together make a *living bio-organism*, once the body engine is placed in very pliant matter such as:
  - Electron gas.
  - Plasma.
  - Living body's complex of body currents.
- Can place an extra mind-engine in a living human body, using the resident body-engine and coupling-engine.

A living mind can be produced and formatted, as can a variant of it (2)

- Can insert the biosystem in an ashtray or in a living body.
- Can insert just the altered mind dynamics (or a fraction of it) in a targeted person's body.

- Switching between original and replacement minds, or between fractions, is called "conversion" or "fractional conversion".
- Conversion dramatically alters individual's behavior, as when multiple personalities switch in a schizophrenic.
- The individual's new behavior is determined by the dynamics designed into the new engine.

Achieving strategic mind war! (1)

- Making one new mind variant or variant of a major mind fraction is difficult and expensive.
- Once made and the sunk costs paid, <u>*unlimited clones*</u> are readily made for a penny a copy.
- Can clone and stockpile many millions of copies, easily and cheaply.

- By paying sunk costs for developing initial fractional engines for a range of behaviors, a vast arsenal of deviant fractional mind engines can be produced and stockpiled for selection, insertion, and switching.

Achieving strategic mind war! (2)

- Decisive strategic mind-war via selected or massive conversions becomes both practical and doable.
- "Sleeper" conversions can be implanted in leaders, key personnel, etc. <u>*in advance*</u> (in an insertion phase).
- Gradual switching generates slowly increasing change in behavior and in internal societal friction.
- Switching massively <u>*all-at-once*</u> generates a sudden and permanent collapse of an entire society into utter chaos.
- The massively switched society violently destroys itself rapidly and completely.

Scenario examples

- President's own converted guards suddenly shoot him.
- Sudden conversion of President, Vice-President, Congress, Judges, Pentagon leaders and high executives paralyzes the nation.
- Police forces suddenly begin attacking civilians. One's own Armed Forces begin destroying each othcr and our own populace and cities.

- The sides of any societal polarization begin vehemently attacking and killing each other.
- Strategic weapons operators "go berserk", change codes, launch on our cities, friends, or indiscriminately.
- Hundreds of other scenarios are possible.

Precursor fields and actions: The secret of "energetics" science (1)

- Force exists in mass and mass systems, *not* in space.
- In space, there is only curvature or distortion regions of spacetime, *with dynamics*.
  - Such dynamic ST regions are force-free "fields" and "potentials" in space.
  - Every field and potential is already a set of energy flows.*
  - The field-in-space is a dynamic "precursor" state, consisting of *energy flow dynamics*.
  - A specific "pattern" of the energy flow dynamics in space and time is called an "engine".
- Engineering the precursor state (*precursor engineering*) is directly engineering nature's fundamental *free EM energy flows and their dynamics*.

* As shown by E.T. Whittaker in 1903 and 1904, without introducing the source charge. Specifically, EM fields and potentials are sets of ongoing EM energy flows, freely streaming from the associated source charge(s) producing the field or potential and its energy.

Precursor fields and actions: The secret of "energetics" science (2)

- Precursor engineering:
  - Basically negentropic: only a little entropy is produced for control.
  - Ongoing interaction of dynamic *precursor engine* with matter produces a directly analogous *force engine* acting in that matter.
  - Controls, directs, switches nature's free energy flows!
  - Extends "gauge freedom" notion, in that *the energy itself is free, and one only pays for a little switching and control of nature's free energy flows*.
- Body and its dynamics are dynamics in 3-space.
- Mind and its dynamics are EM dynamics on the time axis. Can be directly engineered via EM methods based on quantum field theory's time-polarized photon.

About models and modeling (1)

- Choice of fundamental units in one's physics model is *totally arbitrary*.*
- Consider a model with only a single fundamental unit: the joule. Everything else — including length, time, mass, etc. — becomes a function of energy.
- We can even make that fundamental unit a *joule of EM energy*. A priori, energy can be changed in form to any other form. Everything then becomes a *function of EM energy*.
- Thus time has (and is) a special form of EM energy, just as is mass.

* E.g., see discussion of such systems of "natural units" in J.D. Jackson, Classical Electrodynamics, Wiley, NY, 1999 in his Appendix on Units and Dimensions.

About models and modeling (2)

- This extends the time-polarized photon concept in quantum field theory, since all dynamics and functions on the time axis can be considered special *EM energy dynamics and functions*.
- Mind operations and dynamics are on the time-axis (they occur in time, not in space). Hence they also are *special EM energy dynamics*, capable of being directly engineered.
- By engineering time-polarized EM photons, fields, and waves, mind dynamics can be directly produced and engineered.
- Time-energy-EM dynamics in a given mind can also be directly engineered. Another set can be cloned identically, or another set cloned and *altered as desired*.
- Two variations can be made, inserted in a body, and their connection to the body can be switched between them.

Precursor engineering and mind conversion (1)

- A specific set of forces in/on matter is an ongoing *force (material) engine* formed and acting in/on that 3-matter.
- *Producing* it, there is an ongoing interaction of a precursor (nonmaterial) engine and that 3-space matter.
- In the mind realm (time), there are internal patterns, compressions, rarefactions, etc.
  - Compressions of time energy are analogous on the time dimension to "matter", so to speak, in 3-space.
  - The mind interactions can be roughly thought of as analogous to the spatial situation where precursor engines act on "densities of energy called matter" to generate the final "force engines".
- One speaks of "mind energy" and "mind force", to separate thinner intent time actions from the final mind engine that has rotations into 3-space and the body.

Precursor engineering and mind conversion (2)

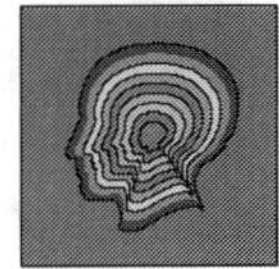

- *Mental intent* acting on denser time creates *mind engines* on the time axis.
- By relativistic 4-space rotation, mind engines also project *coherent EM engines in 3-space virtual state*.
- Negative entropy operation of charges in the brain and nervous system of a body integrates the coherent virtual mind engines into observable EM engines.
- Thus mind engines cohered into observable 3-space engines acting on coupled matter in a body create *intentional behavior in that body*.
- Creation and switching of different precursor mind engines is *conversion*. Switching of fractional precursor mind engines is *fractional conversion.*

Precursor engineering and *nonmaterial* robots (1)

- For every material robot, there exists a force engine for its dynamics, and therefore a precursor (force-free) engine for them. This precursor engine is a *nonmaterial robot* a priori.
- The precursor engine may be produced alone, then inserted into an electron gas, plasma, or other actual physical system to produce the robotics functioning of a material body.
- The precursor engine "nonmaterial robot" can be inserted into distant targeted national facilities, equipment, weapons, etc. and then activated on call, *producing material robots that act on call*.

Precursor engineering and *nonmaterial* robots (2)

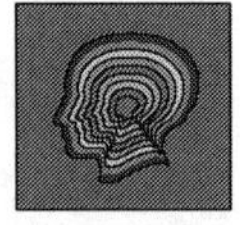

- Start with a precursor engine "nonmaterial robot" and an artificially created mind (dynamic intent engine and corresponding mind engine).
- Coupling of mind (precursor) engine and a precursor robot engine gives a "*living nonmaterial robot*" (a living and dynamic 3-space virtual state engine robot in 4-space).
- The living nonmaterial robot engine set can also be inserted into distant targeted national facilities, equipment, weapons, human bodies, etc. and activated in compliant mass into intelligent functioning.
- *Energetic robots dramatically extend material robotics*.

What this briefing is about

- Background of *energetics* weapons is included, with emphasis on conversion, its science, and its intended use.
- Part of what the FSB/KGB developed as *energetics* superweapons science.
- Three branches, according to what is targeted:
  - *Energetics* (against inert physical systems, targets).
  - *Bioenergetics* (against living biological bodies).
  - *Psychoenergetics* (against living minds and behavior).
- Dramatic tests in 1996, 1997, etc. are detailed.
- In Russia, energetics superweapons science is still highly classified; operated by FSB/KGB, not armed forces.
- West has not developed energetics or defenses against it.

FOREWORD

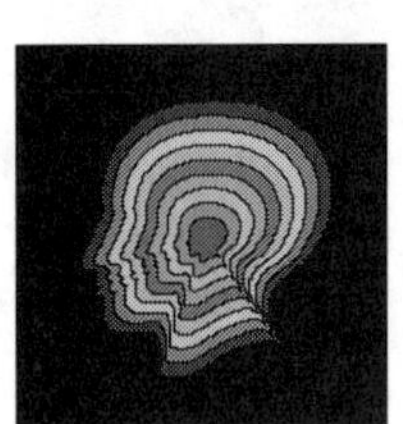

Foreword

Strategic asymmetric warfare plan of the FSB/KGB with the special role of the Yakuza

Foreword (1)

- Multiple strategic WMD threats presently converging to destroy the U.S.
  - Quantum potential weapons
  - EMP and negative energy EMP weapons
  - Scalar interferometry weapons
  - Nuclear weapons inserted
  - Biological warfare threat
  - Bioenergetics threat*
  - Psychoenergetics threat**
- Knockout capability achieved

* Little understood or appreciated by U.S. intelligence (uses electrical engineer analysts).
** Not understood or believed by U.S. intelligence (no scientific analysts even understand it).

Foreword (2)

- Have passed from insertion phase of strategic asymmetric warfare to a 3-year operations phase
- Operational incidents will increase next 2 years inside U.S. while U.S. dragon is increasingly "stretched and bled" overseas economically

- Two years from now, the threats will generate catastrophic economic collapse of the United States, then begin methodically destroying it.
- Will increase damage to centralized electrical power system and destroy it two years hence.

Foreword (3)

- An immediate threat is the psychoenergetics threat (especially personality conversion) augmented by nonmaterial carrier robots.
  - Actual usage starts with isolated incidents and slowly increases essentially without limit.
  - Underway slowly but increasingly at present after recovery delay caused by Israeli destruction of KGB mind-control facilities sometime after mid-1997.
- Main focus of this brief is the conversion threat, with limited background of overall energetics threat.
- U.S. must rapidly achieve and implement counter, or in 2 years probability of our total destruction is very high.

Japanese Yakuza play a special role as FSB/KGB protégés (1)

- Regard themselves as last of the Samurai.
- Often glorified in Japanese movies and media.
- View our defeat of Japan in WW II and our dropping the atomic bomb on Japan as the greatest dishonor in Japanese history.
- In their view, the only way to erase this shame is to destroy every living American.
- In a giant land scandal in Japan, Yakuza ripped off Japanese taxpayer for about $300 billion.
- Penetrate all large Japanese companies and government.
- Allied with old Aum Shinrikyo for technical expertise.

Japanese Yakuza play a special role as FSB/KGB protégés (2)

- FSB/KGB protégés for strategic asymmetric war in the U.S., coordinated by FSB/KGB communist faction.
- Acquired scalar interferometry super-weapons at end of 1989, along with EMP and negative energy EMP aspects, by leasing them from KGB in Russia.
- Can use high altitude EMP or negative energy EMP to knock out centralized power grid at one shot, along with a myriad other electronic systems such as power plants.
- Achieved single-shot destruction capability against U.S., by violently erupting Yellowstone supervolcano at will.

Japanese Yakuza play a special role as FSB/KGB protégés (3)

- Yakuza leased use of scalar interferometers from FSB/KGB in latter 1989.
- Engineered weather over North America.
- Shot down aircraft and missiles, etc.
- Initiated earthquakes and stimulated volcanoes into eruption.
- Protégés for Russian FSB/KGB for internal damage and destruction of the U.S.
- About 1 Jan. 2005, moved into final 3-year operations phase to destroy U.S. Two years to generate incidents, then catastrophic collapse of U.S. economy by destruction of centralized electrical power system etc.
- Then methodically destroy all Americans.

EMP: Expert testimony to Congress*

- *"A nuclear bomb detonated high above the United States could unleash an electromagnetic pulse that would shut down the nation's power grid and, along with it, communications, water supplies and even food transportation."*
- *"If the effect is long-lasting enough, it also could trigger a social collapse that could conceivably cause the deaths of millions of people and, temporarily, push the nation back 100 years..."*

Note: Yakuza scalar interferometers can produce strong EMPs anywhere in and over the U.S. at will.

•As reported by Dee Ann Divis, "Protection not in place for electric WMD," UPI, Mar. 9, 2005, reporting on testimony by Lowell Wood and Peter Pry, respectively member and senior staffer on the Commission to Assess the Threat to the United States from Electromagnetic Pulse Attack. Wood and Pry testified on Mar. 8, 2005 before the Senate Subcommittee on Terrorism, Technology and Homeland Security.

Yakuza achieved U.S. knockout capability as scheduled

- In last quarter of 2004, Yakuza zeroed-in large scalar interferometers on the Yellowstone supervolcano.
- Violent Yellowstone eruption would destroy 70% of North America. Would destroy almost all of the United States.

- Yakuza achieved a one-shot U.S. knockout capability.
- Yellowstone susceptible, sensitive, overdue for eruption.
- Circa beginning of 2005, asymmetric war against U.S. changed from insertion phase to final operations phase.
- For major demonstration at end of insertion phase, Yakuza generated 9.3 seafloor earthquake on Dec. 26, 2004, producing devastating tsunamis in Asia.

Hostile quake-inducing usage is confirmed

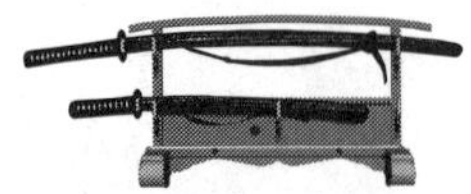

- In 1997 U.S. Secretary of Defense William Cohen* confirmed that such terrorist quake-induction was occurring. He stated:

"Others are engaging even in an eco-type of terrorism whereby they can alter the climate, set off earthquakes, volcanoes remotely through the use of electromagnetic waves... So there are plenty of ingenious minds out there that are at work finding ways in which they can wreak terror upon other nations..."

"It's real, and that's the reason why we have to intensify our efforts, and that's why this is so important."

---

* Quoted from DoD News Briefing, Secretary of Defense William S. Cohen, Q&A at the Conference on Terrorism, Weapons of Mass Destruction, and U.S. Strategy, University of Georgia, Athens, Apr. 28, 1997.

Alliance in progress

- China, Russia, and India alliance growing.
- Energy is a pressing problem for China.
- U.S. now hampered in access to Russian oil.
  - Helps "bleed the U.S. dragon" in preparation for the hidden program to cause catastrophic U.S. economic collapse.
  - Probably ordered by FSB/KGB communist faction.

MBT T-90S

- China "bought into" the Russian oilfields and program.
- China and Russia announced they will hold military maneuvers and exercises in China later this year (2005).

Alliance in progress (2)

- China has 200,000 Chinese in Panama; controls both ends of the Panama Canal. Has access ports in U.S. and this hemisphere.
- China has full complement of energetics superweapons and nuclear weapons.
- China active in North Korean situation. North Korea admits it has nuclear weapons.
- Iran, Syria, Iraq, Afghanistan, Korea are hot spots, will continue to be hot.
- Watch for heat-up of the Taiwan situation, India-Pakistan also. China will take Taiwan by force.
- South China Sea declared as Chinese territorial waters.

Present Operations Phase

- FSB/KGB/Yakuza Operations phase of the asymmetric war is the three years we are now in. The first two years are critical to safely knocking out the U.S.
  - Yakuza has U.S. knockout capability
  - Shot giant 9.3 seafloor quake 12/26/04 generating terrible Asian tsunamis.
- Plan is centrally coordinated by FSB/KGB die-hard communist faction, which retains and operates the Russian superweapons and is the main power in Russia. Yakuza are FSB/KGB protégés.

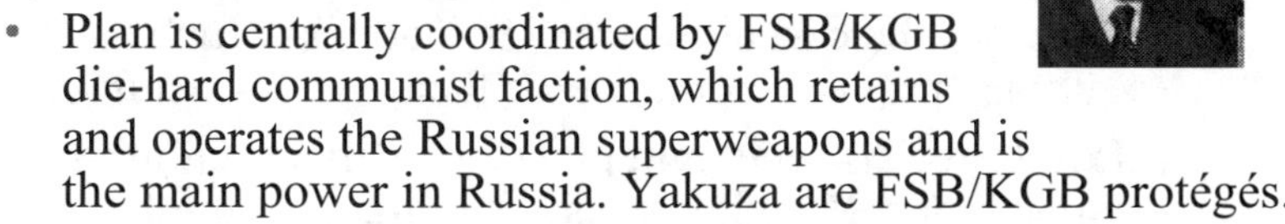

- Putin from younger KGB faction; wishes accommodation with U.S. Not the main power in Russia; follows orders.

Present Operations Phase (2)

- Incidents will begin gradually in the U.S. and escalate. Al Qaida and other terrorists participate as they can.
- Yakuza scalar interferometers gradually destroy U.S. centralized electrical system, some refineries.* Also use EMPs.

- Causes catastrophic U.S. economic collapse. Massive sleeper conversions nullify friendly Israeli quantum potential counters.
- Yakuza methodically destroys prostrate U.S. cities, installations, etc. — fulfilling its mission of killing every living American.
- *If necessary*, Yakuza will erupt the Yellowstone supervolcano to assure total destruction of the U.S.

* E.g., BP refinery explosion in Texas City, TX in latter March 2005 is suspicious. The fires and explosion followed an FBI warning that terrorist attacks might be made on Texas refineries. No one has considered scalar interferometry could be involved.

Bioenergetics aspects of the Operations Phase

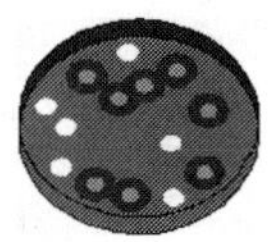

- FSB/KGB will "touch up" American immune systems with quantum potential weapons.
  - "Spreads" immune system's assets to prepare for dozens of expected "shadow diseases".
  - Not much additional assets left for the spread immune system to react to actual pathogens.
- Increases the casualties from BW attacks (such as anthrax sprayed in cities) by a factor of at least three.
- Two terrorists, one light plane, a small agricultural sprayer, and 100 kilograms of anthrax can produce 3 to 9 million casualties in a sprayed city.*

* Due to immune spreading, this is three times the results of the official study done as U.S. Congress, Office of Technology Assessment, Proliferation of Weapons of Mass Destruction: Assessing the Risks, Government Printing Office, Washington, D.C., 1993.

Bioenergetics aspects of the Operations Phase

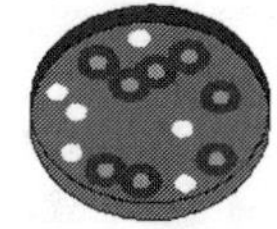

- Anthrax spray attack and other BW attacks by terrorists will thus be unexpectedly devastating.
- The BW threat includes anthrax, smallpox, and many other dread diseases including avian flu and Ebola.

- No truly effective US BW treatment is available for mass casualties.
  - Most casualties will already have symptoms.
  - Will be assessed as <u>triage category 4</u> (black) and just dragged aside to die without treatment.
- Essential decontamination spraying with harsh chemicals will kill thousands of additional citizens.

Devastating use of conversions (1)

- FSB/KGB can and does make *<u>living minds and biological system engines</u>* in the lab.
- Insert a target's mind-clone (with built-in specific behavioral changes) in a targeted individual's body and nervous system, waiting.

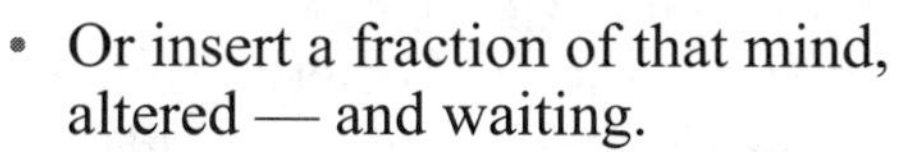

- Or insert a fraction of that mind, altered — and waiting.
- Plant such "sleeper" conversion clones in hundreds of thousands of key persons, awaiting activation.

Devastating use of conversions (2)

- Switch selected inserted minds or fractions at will, from a distance.
- Switch massively. Behavior of masses of leaders and critical persons instantly alters, causing total chaos.
- System begins intensively fighting and destroying itself.
- Produces the ultimate "Manchurian candidate" nightmare.

- ***Also counters the friendly little nation previously saving the U.S. with its quantum potential weapons. It suddenly destroys itself in utter societal chaos.***

General scenario and status (1)

- Yakuza achieves knockout capability against U.S. *(done).*
- Massive sleeper conversions inserted. Counters strategic defenses (*ongoing*).
- Launch operations phase (*operations phase begun*).
- Stretch out U.S. overseas, increasing economic difficulty, "bleed the dragon" until it is hemorrhaging copiously (*begun, will increase*).
- Increase polarization intensities in our society (*begun*).
- BW strikes, in-place nuclear weapons strikes, inside U.S. (*ready*)
- Weather, quakes, volcano eruptions, tsunamies (*set and ready*)

General scenario and status (2)

- Spread immune systems increase casualties (*ready*)
- Destroy centralized U.S. power system (*capability now*)
- Massive conversions in; total chaos (*insertion begun*)
- Catastrophic economic collapse of U.S. occurs (*begun and progressing*)
- QP strike neutralizes U.S. nuclear weapons (*set to go*)
- Yakuza interferometers freed (*weapons poised*)
- Yakuza methodically destroys cities, installations, all living Americans (*ready*)
- If necessary, Yakuza strikes Yellowstone supervolcano, destroying the U.S. *(ready)*

Point and counterpoint (1)

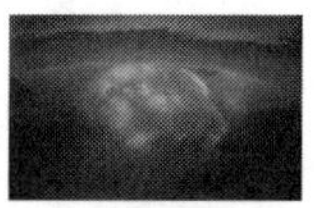

- North Korea announced it already has nukes.
- Will be part of the "disturbances abroad".
- Itself stretched, U.S. may consider arming Japan and South Korea with nuclear weapons.
- If arm Japan, some of the nukes will pass into Yakuza hands. Yakuza undoubtedly working on obtaining nukes now; almost certainly has some.
- Some Soviet-inserted nukes already in the U.S. may be given to Yakuza anyway.
- Having scalar interferometers, Yakuza automatically has EMP and negative energy EMP weapons. Will use such weapons also.

Point and counterpoint (2)

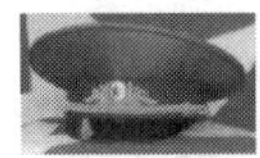

- Insertion phase completed.
- Nuclear weapons inserted, ready.
- BW capabilities inserted, ready.
- Limited conversions are increasing polarizations.
- Massive sleeper conversions inserted or being inserted.
- Yakuza knockout capability achieved against U.S.
- Has knockout capability *now* against U.S. power system.
- Hot spots overseas already heating up.
- Alliance between Russia, China, India.

Point and counterpoint (3)

- Have entered final Operations Phase now.
- Two years until catastrophic U.S. economic collapse and chaos, following destruction of electrical power system, some refineries, etc. by Yakuza scalar interferometers.
- Yakuza then methodically completes destroying U.S. in the third year.

Why the Yakuza?

- Yakuza consider themselves as last of the Samurai.
  - Greatest shame in Japanese history was defeat in WW II and U.S. dropping atom bomb on Japan.
  - Can only erase it by killing every living American man, woman, and child.

- Allied with Aum Shinrikyo, Yakuza leased KGB large scalar interferometers on site in Russia in 1989.
- FSB/KGB uses Yakuza for protégées, to perform the direct "internal" war against the U.S. and its populace.
- U.S. intelligence has no notion of the Yakuza as a formidable force easily capable of destroying U.S.

EXECUTIVE SUMMARY

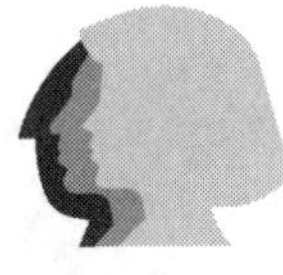

Executive Summary

With accepted threat included

Outline (major sections)

- **Preface, Foreword, Outline and Executive Summary**
- **Background and weapons testing**
- **Move toward first strategic energetics attack**
- **Continued preparations for new attack posture**
- **Second and third attacks scheduled and aborted**
- **World's first subspace robot war**
- **Countdown toward Armageddon**
- **Bioenergetics and its weaponization**
- **Psychoenergetics and its weaponization**
- **Conversion: Mind, behavior, and personality control as primary weapon**
- **Appendix 1: Problems in Western science**
- **Appendix 2: Technical-Theoretic Vulnerability of Network-Centric Warfare**

Weapons of mass destruction (WMD) threat to populace (1)

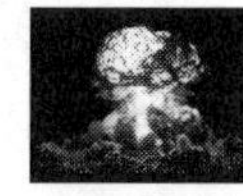

RECOGNIZED TODAY:

- Sabotage and interdiction
- Chemical and biological weapons
- Nuclear materials and weapons
- Powerful explosives
- Assassination, hostages
- Contamination (NBC)
- Shoulder-fired AD missiles
- Terrorism
- Nuclear weapons hidden in our cities (but shunned)

Weapons of mass destruction (WMD) threat to populace (2)

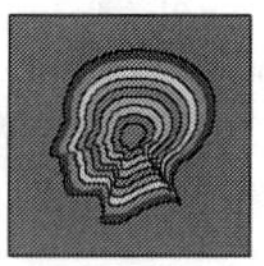

PRESENT BUT UNRECOGNIZED:

- Negative energy EMP weapons
- Quantum potential weapons
- EM bio!ogical warfare
- Scalar EM weapons
- EM disease induction and spreading of immune systems
- Alteration of behavior and emotions
- Alteration of memory and perception
- Control of thought and behavior
- Action at a distance effects
- Causal System Robots (CSRs)
- Massive sleeper conversions

USGS PHOTO

Russia has long had the inserted ability to totally destroy the U.S.

- During Cold War, Soviets inserted nuclear weapons and Spetznaz teams into our major cities and population centers.*
- Spetznaz teams can detonate the nuclear weapons, producing from 100 to 150 million casualties and destroying the U.S.
- Israelis inserted similar teams and nuclear weapons (including thermonuclear weapons) into Russia. Deterred Russians.
- These hidden weapons were known as "dead man fuzing". Gave mutual assured destruction deterrence with its "assured" capability on both sides (the reason the Cold War was "cold").
- Weapons are still on site in U.S., under command of die-hard old Communist faction of KGB (with new FSB name).

---

* Lunev's book gives some of the ways the weapons were inserted. See Stanislav Lunev (with Ira Winkler), Through the Eyes of the Enemy, Regnery, Washington, 1998, p. 22-33. Lunev is our highest ranking GRU defector. Common shipping containers are also handy for inserting.

China is also involved (1)

- China declared South China Sea her own territorial waters
- China will eventually take Taiwan by force
- Has massed missile artillery trained on Taiwan
- Held invasion practice on island near Taiwan

China is also involved (2)

- China has deployed quantum potential weapons
- Deployed fearsome EMP weapons utilizing negative energy rather than positive energy (based on Dirac sea theory)
  - Strike of negative energy extinguishes all electron currents
  - Duds electrical and electron systems instantly
  - Instantly kills struck bodies (zeroes brain, nerve currents)
  - Instant kill of missiles, aircraft, and electronics in nuclear warheads
- Ultimate "death ray"; make large or small, even portable
- Chinese "pinged" this analyst, trying to see how he knew the "cold energy" weapons had been deployed

China is also involved (3)

- After China's invasion practice, in mid-August 2001 a 2-carrier-group U.S. task force maneuvered in South China Sea to caution China
- The new Chinese negative energy EMP weapons were trained on the task force
- Could destroy it within 10 minutes
- Chinese military strike aircraft buzzed the ships and carriers, repeatedly trying to evoke U.S. forces to fire on them. Fortunately U.S. didn't.
- Had we fired, the task force would have been quickly destroyed as test of new negative energy EMP weapons.

Estimate on China (1)

- China given favored nation status and use of two U.S. ports during Clinton administration (move to encourage peaceful accommodation).
- Controls ends of Panama Canal.
- Has inserted some 200,000 Chinese into Panama.
- Eventually will outnumber the Panamanians.
- Panama will become a sort of Chinese province.
- China has other ports in the Western Hemisphere.
- Powerfully driven by escalating energy needs. Increasingly competitive for oil, gas, etc.
- New alliance between Russia, China, and India.

Estimate on China (2)

- Has nuclear weapons, and deployed quantum potential weapons and negative energy EMP weapons, the two most powerful weapons on Earth, other than the robots and conversion. Probed this analyst to see if he knew of the negative energy EMP weapons and deployment.
- Preparing both ways; for accommodation with U.S. and also to be able to destroy the U.S.
- Will take Taiwan by force, initiating conflict with U.S. if necessary. Hence needs dead man fuzing of U.S.
- China has probably already inserted some nuclear weapons, BW agents, and the required teams inside many of our major cities and population centers.

Militant Arab terrorists' asymmetric warfare plan (1)

- Terrorists understand asymmetric warfare and WMD.
- Learned from "dead man fuzing" between Russia and U.S.
  - MAD deters *rational* opponent such as chess players.
  - Does not deter *irrational* opponent.
  - Inserted WMDs provide a knockout capability.
- Intent to achieve strategic knockout capability against U.S.
- Together with Israel, the "Great Satan" is to be destroyed at all costs. WMD, particularly BW and nukes, are required.
- Part of unceasing holy Jihad by radical Moslems.
- Will not end until U.S. is destroyed or until fanatical Islamic groups are nullified.

Militant Arab terrorists' asymmetric warfare plan (2)

IRAN X-55 MSL

Osama bin Laden

IRAN PRESIDENT
Mahmoud Ahmadinejad

- Main objective is to obtain nuclear weapons and clandestinely insert them into U.S. cities; then detonate them.
- 30 to 50 inserted warheads needed.
- Al Qaeda may have 7 to 15 nuclear warheads already inserted (10 and 100 KT) in 7 to 15 U.S. cities.
- May have others for additional insertion.
- Supportive Arab states such as Iran are intent on obtaining nuclear weapons capability of their own.
- Russians are assisting them to build nuclear facilities.

Mechanism to obtain nukes (1)

- Iran and terrorist-backing states seek nuclear power plants.
- Iran is constructing several such plants; one hidden underground.
- One in particular, being constructed by the Russians, will produce weapons grade plutonium.
- Iran intends to become major nuclear power.
- In 2004 Iran and Russia announced this power plant has been completed

IRAN NUCLEAR POWERPLANT

* Russia now contracted to help Iran finish, and to furnish fuel.

Mechanism to obtain nukes (2)

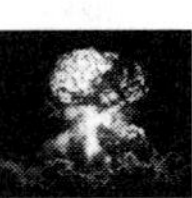

- The Iranian nuclear plant could be producing in perhaps a year, since Russia is helping and will furnish fuel.
- Yearly it will produce weapons grade plutonium for 20 to 50 nukes.
- Separation of the plutonium is straightforward.
- In about 2 years, up to 30 nuclear weapons may be available, and more each year thereafter, from that plant.
- Other nukes will be available from weapons grade material of other plants.
- Negotiations to avoid this continue, with little avail.

IRAN NUCLEAR POWERPLANT

Mechanism to obtain nukes (3)

- North Korea has some nuclear weapons; is producing nuclear plants. Al Qaeda has a few purchased nuclear weapons.
- Several other Arab nations backing terrorists considering same.
- About 2 to 3 years from now, Al Qaeda and other organized terrorist organizations will have their nukes inserted. *May be at half-capability now*.
  - Can then utterly destroy the United States.
  - Have purchased at least a few nuclear weapons.
- Have obtained and inserted BW including anthrax.

IRAN NUCLEAR POWERPLANT

On Al Qaeda plans:

- Insertion of nuclear weapons relatively easy and safe.
  - Used method proven by large drug dealers.
  - 98% of all standardized shipping containers were never inspected anywhere in the world.*
- Hid weapons in standard shipping containers.
  - Formerly high probability of getting through undetected.
  - U.S. trucking industry and railroads will load the containers at the port and transport them to the waiting terrorist teams on site in U.S.
- Clandestine transport across unguarded and remote wild U.S. borders by teams, and insertion by sea.
- High probability of success in insertion. U.S. borders unsealed; insertion across borders readily possible.
- U.S. has deployed nuclear detectors in U.S. cities, and is acting to prevent or hamper new WMD insertions.

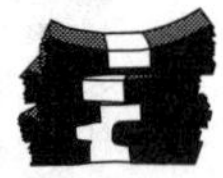

Terrorists understand economic warfare (1)

- Quoting Osama bin Laden* after 9/11:

- *"Al Qaeda spent $500,000 on the [September 11 attacks], while America, in the incident and its aftermath, lost — according to the lowest estimate — more than $500 billion. Meaning that every dollar of al Qaeda defeated a million dollars, by the permission of Allah!"*

- He declared his policy is "*bleeding America to the point of bankruptcy.*" Joked about U.S. "*economic deficit ... estimated to total more than a trillion dollars..*"

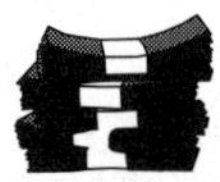

Terrorists understand economic warfare (2)

- Quoting Niall Ferguson*:

"A doomsday scenario is plausible. ..another, bigger September 11 is quite likely... a crisis over Taiwan would send shockwaves through the international system ... revolutionary regime change in Saudi Arabia would shake the world ... detonation of a nuclear device in London would dwarf the assassination of Archduke Ferdinand as an act of terrorism... Like the passengers who boarded the Lusitania, all we know is that we may conceivably sink. Still we sail."

---

* Niall Ferguson, "Sinking Globalization," Foreign Affairs, 84(2), Mar./Apr. 2005, p. 76-77. He is Professor of History at Harvard University, Senior Fellow at the Hoover Institution, Stanford University, and Senior Research Fellow of Jesus College, Oxford Univ.

 The growing U.S. financial problem* 

- Huge spending ongoing: "Guns, butter, tax cuts".

  *"The federal government's obligations, current liabilities, and unfunded fiscal commitments are over $43 trillion and rising...."**

- As Zuckerman points out:
  - *"The estimated net worth of American families is slightly over $47 trillion, and nearly all of it — 90 percent — would be needed to cover government's current obligations."**
  - *"According to the GAO, it would take real double-digit growth over the next 75 years to pay off our current debt — an impossible task..."**

---

* David Walker, head of the Government Accountability Office.

** Mortimer B. Zuckerman, "The case of the 12 zeros," U.S. News & World Report, Mar. 21, 2005.Zuckerman is editor in chief of the news magazine.

Refinery explosion in Texas City, TX in March, 2005

- Could be scalar interferometry-initiated, in which case U.S. agencies would conclude it's not terrorism (and they did).
- Agencies would not recognize scalar interferometry use.
- E.g., FBI pulled back its anti-terrorism investigators. Quoting FBI spokesman Al Tribble*:

"There's no criminal investigation as it pertains to terrorism at this time..." "We're convinced through our bomb technicians, our personnel that were sent down there, the authorities that are on-site, and the BP officials that there are no reports of any breaches of security . There's been no suspicious activity reported. The blast characteristics aren't consistent with any sort of terrorist-type of threat or activity. So we pulled our people back."

- There have been no new U.S. refineries built in a quarter of a century. What we have are running at 96% capacity, barely able to "keep up" with demand.

---

* (McVicker/Hedges, Houston Chronicle).

Success always attracts others

- Russian FSB/KGB faction under die-hard Communists controlling the inserted Russian WMD weapons and super-Weapons; may be unable to resist the opportunity.
- China may be unable to resist the opportunity.
- Yakuza/Aum Shinrikyo <u>*are*</u> actively participating.
- Hence there is a growing, loose but immense attack consortium of terrorists, Russia, China, Yakuza/Aum Shinrikyo, North Korea, etc.
- Any major terrorist nuclear attack on the U.S. may draw consortium into full massive attacks on the U.S.
- Probability of destruction of the U.S. is very high; critical point by two years from now *(now* being Feb 2005).

Update: Yakuza role in asymmetric war on U.S.

- The Yakuza is a major protégé of the FSB/KGB; is coordinated and to be used against the internal U.S.
  - Yakuza regard themselves as last of the Samurai.
  - Greatest shame in Japanese history was our WW II defeat of Japan and dropping atom bomb on Japan.
  - Stain can only be removed by the total destruction of every American man, woman, and child.
- Operations phase: Gradual economic bleed and destruction, particularly of U.S. electrical power system.
  - Centralized system will be destroyed in two years.
  - Simple since scalar interferometers are available.
  - Evokes catastrophic economic collapse of the U.S.
  - Western European economy also collapses in response.
- Yakuza will then destroy U.S. cities and populace at will.

Bottom line: WW III in progress. "Peacetime" was insertion phase (1)

- In 2 years or less, with a few more nukes inserted in U.S. cities Al Qaeda will also achieve ability to effectively destroy the United States.
- Yakuza already achieved U.S.-destroying capability with scalar interferometers registered on Yellowstone supervolcano and other volcanoes.
- This is WW III. "Peace" is totally eliminated. There are two main phases in this asymmetric war:
  - *Insertion phase*, when WMD are being inserted. Lasted 20 years. Mistakenly called "peace".
  - *Operations phase*, when WMD are unleashed.
- We are now in the Operations Phase vis a vis the FSB/KGB, Yakuza, and Al Qaeda.

Bottom line: WW III in progress. "Peacetime" was insertion phase (2)

- Negotiations, statesmanship, UN debates necessary, but are only coverage for ongoing insertion phase and operations phase. We must also:
  - Decisively defeat future WMD insertions.
  - Nullify previous WMD insertions.
  - Unceasingly attack terrorist elements.
  - Develop Prioré-type effective EM treatment of mass casualties at highest possible national priority.
- U. S. has deployed sensitive nuclear detectors in our cities and population centers, is building a homeland defense, and has taken other measures such as a massive reordering and improvement of intelligence.

Terrorists' backup WMD capability (1)

- WMD such as smallpox, anthrax, dirty nuclear explosions etc. can severely damage or destroy the United States.
- BW agents made by many nations are available. Insertion continues.
- Anthrax spores available for years on Voz Island* in the Aral Sea. Those who wanted anthrax, got it easily. Already inserted into the U.S.**
- Smallpox and BW Soviet scientists were available for buy and hire, and multiple terrorist-backing nations bought and hired them.

* Vozrozhdeniye Island is located in the Aral Sea between Kazakhstan and Uzbekistan. Hundreds of tons of biowarfare agents, etc. were dumped there by the Russians.

** Witness the high grade anthrax mailings (a test, not a strike) after 9/11.

Terrorists' backup WMD capability (2)

- Significant insertion in U.S. already accomplished, especially anthrax.*
- A single professional anthrax attack on Washington DC, sprayed from a light aircraft, will produce 1-3 million casualties.**
- If smallpox is unleashed in a single major city, it will eventually kill some two billion persons worldwide — nearly 1/3 the human population.
- Will be unleashed — matter of <u>*when*</u>, not <u>*if*</u>.
- The inserted nukes and dirty weapons will also be unleashed.

* Witness the high grade anthrax mailings (a test, not a strike) after 9/11.

** KGB/FSB's immune system spreading increases these BW casualties by a factor of 3.

Additional threats (1)

 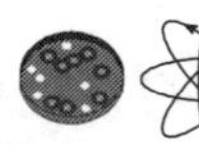 

- Russian nukes and Spetznaz teams already inserted in U.S. cities, ready.
- Can destroy the U.S. at any time.
- Originally deterred by Israeli nukes inserted in Russian cities.
- Chinese probably inserted or inserting nukes, perhaps BW agents and teams also.
- China planning for either accommodation or attack.
- Yakuza inside U.S., probable significant insertions, have scalar interferometers and EMP weapons, both long range and portable.

Additional threats (2)

- Yakuza building small, relatively portable scalar interferometer weapons.
- Inserting them with teams, to strike nuclear power plants, refineries, and other critical targets.
- Already mans long-range interferometers on-site in Russia.
- Strongly engineers weather over North America, occasional aircraft shootdown, etc.
- Has negative energy EMP weapons also to attack U.S. cities, nuclear power plants, refineries, grid, etc.
- *Directly generated the 9.3 earthquake causing the recent terrible tsunamis in Asia*.

Additional threats (3)  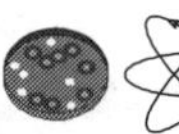 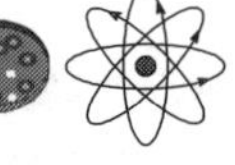

- Castro has inserted 10,000 guerrillas in U.S. over the decades. Perhaps 5,000 are still viable assets for sabotage and asymmetric warfare.
- Miscellaneous BW, chemical, and guerrilla threats from other foreign assets inserted by other nations (Iran, Syria, Libya, etc.).

- U.S. borders still essentially open; insertion still easy and continues.
- Most standard shipping containers still unchecked.

Additional threats (4)  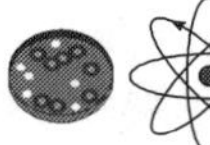 

- Brazil is an another unknown quantity; has scalar interferometers and quantum potential weapons.
- Met with FSB/KGB in Iceland and signed nonaggression pact. Russians have no intention of honoring it.
- In any major strategic strike on the U.S., FSB/KGB must destroy Israel and Brazil due to their QP weapons.
- Brazil hired the German scientific leadership of Russia's scalar interferometry development (by the German radar team) after WW II, once the German team was released by the USSR. This initiated Brazil's superweapon program.
- David Bohm's visit to Brazil allowed picking his brain on quantum potentials. Brazil then developed QP weapons.

Slyness of KGB plans: The chessplayer outlook pervades

- *"Each side secretly develops new means of warfare in order to employ them unexpectedly. History knows many examples how the employment of a new weapon initially gave considerable success because the enemy, caught unawares and not knowing the combat capabilities of the weapon, was for some time incapable of effective counteraction."* [1]
- *"Of particular importance is basic research aimed at discovering still unknown attributes of matter, phenomena, and the laws of nature, and developing new methods for their study and use to reinforce the state's defense capability."* [2]

---

1. V. Yo Savkin, The Basic Principles of Operational Art and Tactics, Moscow, 1972.
2. Portion of a book by Marshall Grechko, deleted from its English translation by specific request of the USSR.

Strengths on the side of the U.S.

- Large strategic nuclear capability.
- Israel has powerful quantum potential weapons, deters many scenarios.
  - Has repeatedly rescued the U.S. from planned FSB/KGB strategic attacks.
  - Has inserted nuclear weapons inside Russian cities, population centers, and key areas.
  - Has at least a limited capability against mind conversion and psychoenergetics.
  - Can defend against precursor robots.
- U.S. believed to have some secret weapons of its own; exact types and details unknown.

Main weaknesses of U.S. could be solved by vigorous action (1)

- Americans do not react to slowly increasing threats.
- Public and officials still not educated as to what is really going on. Other than Cohen's statement, officialdom has been resoundingly silent on superweapons, their development and testing, etc. Intelligence analysis still using puerile old CEM/EE model.
- Historically, American preparation occurs *after* sudden attack has *already occurred* (reacting to the cobra's strike).
- Borders need to be sealed; standard shipping containers need to be checked.
- Develop mass casualties EM healing device (Prioré type).
- Quickly develop and deploy fuel-free electrical power systems taking energy from the seething vacuum.

Main weaknesses of U.S. could be solved by vigorous action (2)

- Predominance of archaic, error-riddled classical electromagnetics (CEM) and electrical engineering discards:
    - Precursor engineering and nonmaterial robotics.
    - COP > 1.0 electrical power systems "fueled" by vacuum energy.
    - Engineering understanding of mind, mind dynamics.
    - Understanding of "electromagnetic" induction of diseases, etc.
    - Understanding of psychoenergetics.
    - Production of living minds and living systems.
- Combination places U.S. at *grave strategic disadvantage*.

---

* T.E. Bearden, "Errors and Omissions in the CEM/EE Model", at http://www.cheniere.org/techpapers/

Energetics: Precursor engineering (1)

- Present CEM/EE falsely assumes:
  - Force fields in space (exist only in matter).
  - A material ether (falsified since 1887).
  - A flat spacetime (falsified since 1916).
  - That EM fields, potentials, and energy are freely created out of nothing at all, by source charges — totally in violation of conservation of energy law.
  - That static EM fields and potentials are *frozen.* Instead, they are nonequilibrium steady-state sets of ongoing dynamic, free flows of EM energy.
- Any amount of force or energy can be collected from a given "static" field or potential, if one has sufficient collecting charge, by $F = Eq$; $W = \phi q$.

Energetics: Precursor engineering (2)

- Present EM circuits are made Lorentz-symmetric, by wiring external source of potential into its own external circuit, while current is flowing, as a load to be destroyed.
- Professors and texts do not calculate the *force-free EM energy field as it exists in space as a precursor of the EM-field of force in matter* . None ever have. Ongoing precursor field interaction with charge gives force fields.
- *EM energy flow is already free*. Force-free EM energy field in space can be engineered by paying a little for switching and control, first making a precursor EM engine with desired dynamics. One need not pay for the energy itself since *all EM energy is in freely flowing streams already, pouring from source charges and dipoles*.

Energetics: Precursor engineering (3)

- There is not now, and there never has been, a single EE department, professor, or textbook that teaches calculation of the EM field in space. Instead, they teach calculation of its *force field effect* in charged matter, created by ongoing interaction of that real EM field with that charged matter.
- There is not now, and there never has been, a single EE department, professor, or textbook that teaches (or knows) how an electrical power system is really powered. It *isn't* powered by cranking the shaft of the generator!
- So long as our scientific community ignores the source charge problem (swept under the rug for a century), we will remain at the mercy of those who *did* correct the flawed CEM/EE model, *did* calculate force-free EM fields in space, and *did* develop precursor engineering.

The "New" Strategic Problem (1)

- *Nonmaterial* energetics robots (self-powering precursor engines) carrying *specific spacetime curvature engines (weapons engines)* have been fully developed and deployed under the FSB/KGB's die-hard Communist faction.

U.S. SCIENCE

- Totally new form of asymmetric war: *Nonmaterial fighting systems*.
    - First war already fought (few days, 1999-2000).
    - Small non-Western nation destroyed robot armada of millions of such robots inside U.S. *in subspace interior to normal EM field signal envelopes*.
- West is 50 years behind, has no defenses against such. Scientific community responsible for U.S. vulnerability.

The "New" Strategic Problem (2)

- Can be superior to all other forms of combat.
- Economics dirt cheap (penny a system after sunk costs paid).
- Effectiveness is complete against defenseless targets.
- Targeted Western nations vulnerable and defenseless.
- Western science is woefully inadequate and completely unaware of this technology; has slept for 50 years, intends to keep on sleeping.
- CIA, NSA, DIA, NSC, NAS, NSF, think tanks, etc. appear essentially unaware of energetics precursor robots technology and development.
- Largely unaware of conversion science and threat. Use standard CEM/EE instead of energetics.

Characteristics of nonmaterial energetics (precursor) robots (1)*

- For every physical system and its functions, there is a corresponding set or "template" of spacetime (ST) curvatures (a set of precursor energy field dynamics), acting on the charges and masses to generate all system forces and their actions.

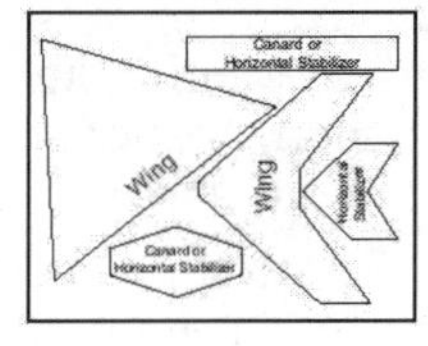

- EM fields and potentials in space are force-free energy fields and specialized curvatures of spacetime. A *template* of desired ST curvatures and their dynamics can be formed separately, as a harmonic set of precursor bidirectional EM LW wavepairs and their dynamics.

---

* Also referred to as "causal system robots".

** Spacetime curvature (energy) may also be taken as change of virtual particle flux of vacuum.

Characteristics of nonmaterial energetics (precursor) robots (2)*

- Precursor robot is a template of ST curvature dynamics.
- It is nonmaterial and force-free.
- When it interacts with charged matter, the ongoing interaction produces force dynamics in that charged matter.

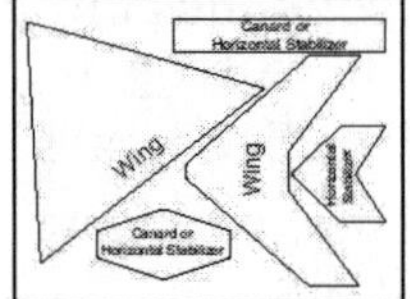

- Functions — communication, weapons, surveillance, weapons functions, etc. — can be "built-in".
- The "robot" is then commanded via EM longitudinal wave LW communications systems.
- An entire robot armada of millions of such robots can be assembled and controlled in distant battle.

---

* Also referred to as "causal system robots".

** Spacetime curvature (energy) is also change of virtual particle flux of vacuum.

Characteristics of nonmaterial energetics (precursor) robots (3)

- <u>*Spacetime curvature is energy.*</u>** Since ST curvature is energetic, the template/robot is a self-powered system, hence a nonmaterial robot fighting system once it interacts with matter.
- As weapons it uses <u>*precursor energy engines*</u>, not the usual material engines.
- Interaction of precursor engines in matter produces the desired real fighting engine effects. Almost any physical effect can be generated.
- Precursor engine robots easily travel through earth, ocean, space, since ordinary EM fields, potentials, and waves are bundles of LWs and superhighways.
- First making *one* robot is expensive and difficult; *next million copies* are 1 cent each.

Implications of "new" warfare (1)

- Demise of the U.S. in 2-3 years is highly probable unless an immediate Manhattan Project is launched under direct Presidential Directive at highest priority and speed.

- Nonmaterial robots and switchable sleeper conversions are presently two of the greatest threats, of the several rapidly converging.
- But the U.S. will likely be struck by all the threats.
- Developing EM treatment of mass casualties is absolutely necessary if to avoid most (being Triage Category 4) just set aside and allowed to die.

Implications of "new" warfare (2)

- To have a survival chance, greatest effort, speed, and closest possible selection of scientists are required.
- Close coordination with friendly foreign weapons scientists at the highest level is imperative. 
- Some nations do have some counter-measures and destroyed the first nonmaterial robot armada attacking in 1999-2000 after being alerted by this analyst and three close colleagues. U.S. has nothing.
- One, possibly two nations may have limited capability against conversion technology.

Conversion: Matured mind control threat capabilities (1)

- Secretly entrain or enslave leaders and personnel at any and all levels
- Install secret multiple personalities
- Switch their personalities at will
- Easily defeat detection systems
- Penetrate all compartments
- Compromise all information
- Manipulate and increase polarization intensities

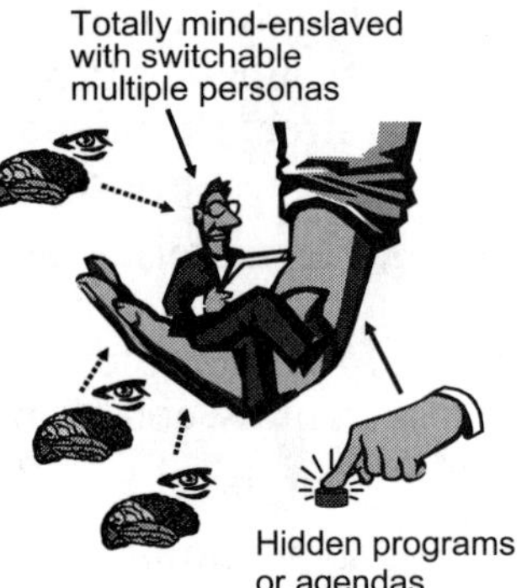

Conversion: Matured mind control threat capabilities (2)

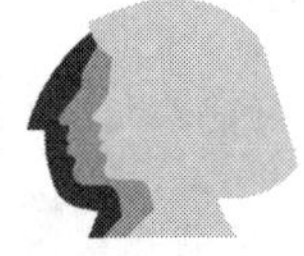

- Carrier robots with mind-engines access every part of the entire nation
- Augments and completes carrier robot forces with bioenergetics engines and energetics engines
- Because of ease of reproduction (penny each), may be cheaper than any other form of combat
- Possible to subvert a government or culture, armies, organizations, nation

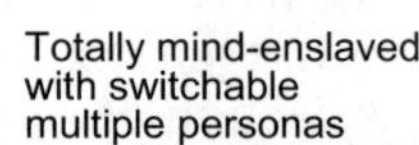

Conventional secrecy is compromised

- "Secure phone" to one's ear with head in the EM field is a direct pipeline into one's entire mind, memory, knowledge, behavior.
- Any signal path into a computer etc. is a pipeline also.
- Psychoenergetics technology can "search, find, and download" the information rapidly.
  - Pass inside EM fields in matter, thus through walls, safes, vaults, mountains, ocean, earth, etc.
  - Pass through encryption device to raw input information.
  - Scan, copy, download, and transmit the information.
  - Penetrate any meeting, vault, sweep processes, etc.
- Limited only by the ability to process the enormous amount of information gathered.

Nonmaterial robot capabilities

- Nonmaterial (precursor energy field) robots with weapons can attack, control, disrupt, destroy
  - Any and all inert matter systems (energetics carrier robots).
  - Any and all living biological systems (bioenergetics carrier robots).
  - Minds of selected personnel (psychoenergetics carrier robots).
- Robotic systems can self-install and reside inside matter or EM field, using the charges of that matter or energy of that field to form a functioning "physical" body.
  - Becomes a mine system designed against any target.
  - Its auxiliary robots can be weapons against any and all targets.
- Ultimate "mine warfare". Segments of (or entire) physical environment of targeted nation can be filled with deadly weaponry unceasingly attacking targets.

To understand the robots, new concepts are necessary (1)

- Infolded interior structuring of ordinary EM fields, potentials, and waves as biwave pairs. Normal EM is just "envelopes".*
- Infolded longitudinal EM waves produce oscillating energy densities (curvatures) of spacetime precursors (basis for engines).
- "Engine" concept (curvatures of spacetime functioning as a system to perform any set of operations upon and in matter desired).
- Precursor engine interacts with mass to produce a real, physical force system (engine) operating on that mass

* Basis has been in physics since two papers by Whittaker in 1903 and 1904.

To understand the robots, new concepts are necessary (2)

- "Mind engine" concept: Electrodynamics based on scalar EM waves on the time axis, using scalar photons from quantum field theory coupled to longitudinal photons, so that couplets are observed as instantaneous scalar potentials (voltages).
- "Nonmaterial robot" system concept as extension of engine and mind-engine concepts. Robots are by definition *self-powering.*
- Carrier robots infold numbers of weapon/action robots within their own interior electrodynamics.

Afterword: Examples of inanity of electrical engineering model (1)

- Has not solved its "source charge problem" in over 100 years. *Solution summarized in this brief; basis in physics since 1957 award of Nobel Prize to Lee and Yang.*
- *Still assumes material ether* (and force fields in vacuum) a century after falsified.
- Assumes all EM fields and potentials and their energy are freely created out of nothing at all by source charges.
- Assumes "static" field or "static" potential is frozen. Each was proven to be a set of inexhaustible, ongoing EM energy flows, a century ago, by E. T. Whittaker.
- Doesn't know or teach what really powers an electrical circuit or device. Not a single EE department, professor, or text ever has. NAS, NSF, NAE, universities, scientific societies etc. not acting to correct these known errors.

Afterword: Examples of inanity of electrical engineering model (2)

- Does not calculate the actual EM field (force-free precursor, which is just a change in curved spacetime or in virtual particle flux of vacuum); never has. Only calculates the *effect in charged matter*, of ongoing interaction of EM field with q. That "effect in matter" is a force field, but it is *not* the EM field that exists in space prior to interaction. *Mass is a component of force!*
- Directly responsible for energy crisis and dependence on foreign oil, and for present total vulnerability of U.S.
- Assumes inert vacuum (false since 1930) and flat spacetime (false since 1916).
- Discards Heaviside energy flow component (up to a trillion times as much flow as in Poynting flow).

Afterword: Examples of inanity of electrical engineering model (3)

- Arbitrarily eliminates COP energy-from-the-vacuum (EFTV) systems, by using circuits that self-enforce Lorentz symmetry (forward emf and back emf forced to be equal).
- That allowed use of simpler (Lorentz-invariant) equations to describe the operation of such "self-crippling" circuits.
- Discarded all permissible COP>1.0 Maxwellian systems.
- As a simple analogy, it destroys — so to speak — electrical heat pumps. Normally a heat pump receives and uses excess energy from the environment, thus enabling COP>1.0 and also self-powering. Instead, symmetrized EM only permits the EM "heat pump" to be on resistance heating, never using an extra joule of EM energy from the environment.
- This is particularly inane since all EM energy occurs in continuing free flows, *never* as so many rocks in a rockpile.

Typical examples of scientific dogma and suppression (1)

- Savaged Meyer on energy conservation.
- Savaged Wegener on continental drift.
- Savaged pioneers of ultrawideband radar.
- Savaged Becker on revolutionary EM healing.
- Savaged Ovshinsky on amorphous semiconductors.
- Savaged cold fusion (600 successful experiments).
- Savages free energy research; any EE department can build a self-powering Takahashi magnetic engine.
- Stopped Fogal technology for superluminal communication and removing noise from digital signals/photos.
- Responsible for energy crisis and untold human suffering.

Typical examples of scientific dogma and suppression (2)

- Ruthlessly suppressed revolutionary electro-healing (unwittingly using *precursor disease antiengines*) by French scientists on the Prioré team.
- Ignored attempt in 1998 to get U.S. to develop it for quick mass casualty treatment and curing. Most BW casualties now will be Triage Category 4, set aside and allowed to die. *Scientific community is responsible*.
- Did not solve four decades of EM induction of diseases and health changes in personnel in Moscow Embassy by precursor field "disease engines".
- Most scientists think we are "meat computers". Have no notion of time-as-energy, scalar photon, or thought and mind operations as time-axis EM dynamics.

Additional observation

- Key persons who could engender understanding of the threat science are viciously attacked and vilified.
- Agents and agents of influence involved, per gaming and disinformation techniques. Murder also used.
- Dramatically reduces probability that U.S. government and scientists wake up to what is ongoing.
- Performed by U.S. cartels and also FSB/KGB.
- For KGB, it is part of the overall attack support plan* to insure that the U.S. sleeps on, preoccupied with Afghanistan, Iraq, Iran, North Korea, oil, reactors, etc.

* An important strategic attack plan will always have an accompanying deception plan. The more important the attack, the more elaborate the deception plan. Obviously the FSB/KGB has a comprehensive (and very successful) strategic deception plan.

A famous quote:

"I know not with what weapons World War III will be fought, but World War IV will be fought with sticks and stones."

Albert Einstein.

- The weapons used in the ongoing WW III are indeed strange and previously unknown in the West.
- Scientifically we have been asleep for decades, it seems, as to the nature of these eerie weapons.
- We must react swiftly, for there is little action time left.

Conclusions (1)

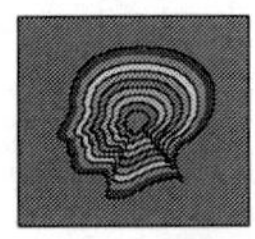

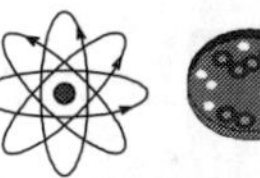

- World War III is in full progress.
- Multiple total destruction plans against the United States now implemented and converging.

- Success probability very high (maybe 98%).
  - Inserted nukes and BW; scalar interferometers, QP weapons; negative energy EMP; robots; mind control; and conversions.
  - U.S. scientists unaware; few defenses.
  - Time is very short — less than 3 years; and two year critical period with extensive conversion strikes.

Conclusions (2)

- U.S. is unacceptably vulnerable:
  - Appalling age and flaws of EM models and electrical engineering.
  - Naiveté, dogma, and cartels in scientific community, industry, government, medical community, etc.
  - Intelligence community and political community unaware; have inadequate scientific advice. Thus ineffective.
  - Nation essentially unprepared for the type war ongoing.
- Main decision makers lack knowledge of eerie new weapons technology and how to defend against it.
- Cannot even recognize nature of an attack (Button, Svoboda, Hess, Thresher, Gander, Challenger, Titan, TWA-800, Yellowstone, Asian tsunamis, etc.).

Conclusions (3)

- Unless galvanized by strongest possible Presidential action and crash project to develop counters, U.S. will likely be destroyed in 3 years or less.*
- Unless forced in strongest manner, the organized U.S. scientific community will not only fiddle while Rome burns, but it will also assure that Rome burns completely.

U.S. SCIENCE

* Now in 3-year Operations Phase. At end of 2 years is the decisive point.

Desperate U.S. needs

- Presidential Decision Directive in strongest terms.
- Force scientific community to correct CEM/EE model.
- Expunge 1% evil entourages in national labs, etc.
- Countermeasures against conversion, precursor robots.
- U.S. quantum potential weapons, precursor robots.
- Mind-engine production capability.
- Fuel-free energy-from-the-vacuum power systems.
- Portable Prioré-type medical treatment units.
- Sealing of borders and inspection of cargo containers.
- Briefing of Congressional leaders and decision makers, mayors, governors, legislatures, police forces, etc.
- Tailored briefing of American populace and news media.

(SLIDE BRIEFING CONTINUED)

PART B: Background and Weapons Testing

BACKGROUND AND WEAPONS TESTING

Background and weapons testing

Russian and Western science compared

Section contents

- **Russian science**
  - **Energetics**
  - **Bioenergetics**
  - **Psychoenergetics**
  - **Engines and robots**
  - **Carrier robots**
- **Western science**
  - **Serious flaws**
  - **Electrodynamics**
  - **Electrical engineering**

Stalin's dictum in 1945 removed any resistance from his scientific community

Assembled Academy

Stalin

"The destiny of communism has been frustrated by the U.S. development of the atomic bomb. That is not the last great technical breakthrough. The next one will be Soviet! You will provide me with that great new technical breakthrough at all speed, or I will have your heads. Do I make myself clear, comrades?"

Response to Stalin's iron dictum

- Soviet Academy of Science desperately sought anything anomalous and not followed up, which could be weaponized for a new "great weapon".
- Search included mind, living bodies, inert matter areas.
- Reviewed, corrected scientific models, e.g., CEM/EE.
- Hauled in copies of scientific literature journals of the West, from Day 1, to analysis institutes by shiploads.
- One institute alone had some 2,000 Ph.D.'s and their staffs (translators, secretaries, etc.).
  - Reviewed each page of each journal issue. Selected interesting items previously not followed up.
  - These reviewed by best nonlinear scientists in the world.
- Energetics born in 3 to 4 years; used to design weapons.

Energetics: Result of this great Russian effort

- Immediately focused on everything important that had not been followed up in science.
- Kept this attitude permanently; researched most promising areas, using new scientific models.
- Effort named _energetics_; has three targeting categories:
  - _Energetics:_ Inert matter and EM fields, potentials, and waves.
  - _Bioenergetics:_ Living biological matter with biofields, biopotentials, and biowaves.
  - _Psychoenergetics:_ Mind with its scalar (time-polarized) electrodynamics).
- Expanded QFT's scalar and longitudinal photons; couplet observed as scalar potential (voltage).

Targets of the three branches of energetics weapons

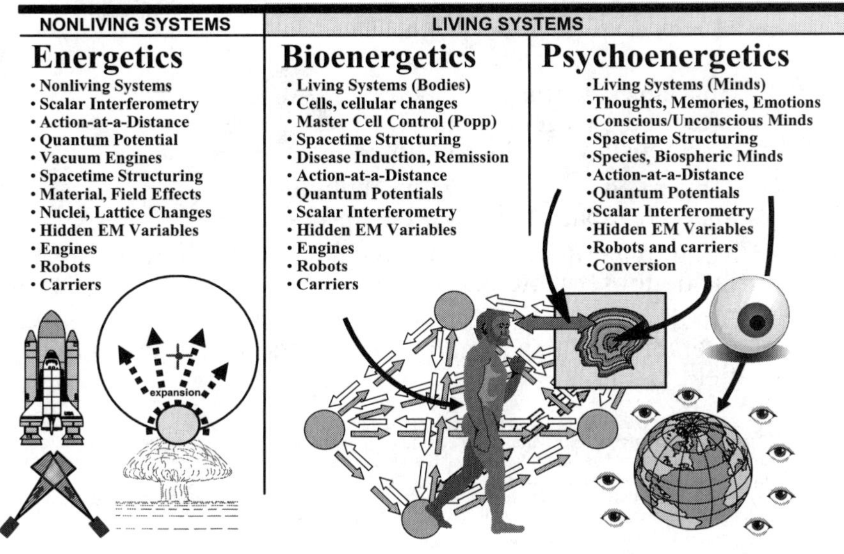

Origin of term "energetics"

WILLIAM RANKINE

- *Energetics* term recognizes energy as the primary thing in the universe.
  - Rankine first used it.
  - G. Helm revived it, separated it into an *intensity* factor and an *extensity* factor.
  - Ostwald in 1893, 1895, and 1901 raised the energetics concept to the forefront.
- In general relativity terms, curvature of spacetime is essentially the same as "energy" or "energy in motion".
- All forces are made by *ongoing* interaction of ST curvature* with matter (mass-energy).

* In particle physics, substitute "change in VPF of the vacuum" for "ST curvature".

Extending energetics

- Russians scientists extended energetics into n-dimensions, including time.

- "Observed" physical reality includes mass and space but not time, and time is only inferred.
- Mind dynamics are electromagnetic and originate on the nonobservable* time axis.
- Mind dynamics are EM time-energy dynamics, interacting with a mind/body connection per quantum field theory.

•Time is not an observable, even in theory. Coherent time dynamics, however, are input as 3-space virtual state projections to the charges in the brain and nervous system. The charges integrate coherent integrations of virtual 3-space projections from the mind's intent dynamics, outputting observable photons which are *EM observables* correlated to *coherent intent* time dynamics. E.g., that is how the body-mind coupling is able to produce output observable EM nerve spikes and signals coherent with mental (time domain) intent of the mind.

Mind-to-body coupling mechanism

- Time is not an observable, hence mind is not observable.
- Mind changes on time axis project virtual state intersections in 3-space, into the coupled nervous system.
  - *Charges* in body system provide coherent integration of virtual state changes into observable photons.
  - Intent changes thus integrated into coherent observable EM, with small integrating delay.
  - Occurs in "spike" areas (in dendrite discharges).
- Coherent integration produces intent-coherent EM discharges in body's nervous system.
- Servomechanism control theory then applies in body.

Body-to-mind coupling mechanism

- Body changes reflect virtual state changes onto time axis, into coupled mind dynamics.
- Time-energy virtual changes are also coherently integrated by the 4-space charges in body.
- Produces coherent "perceivable" mind changes coherent to body responses (after small delay).
- Mind perceives these changes, compares, readjusts, etc.
- Delay in total loop gives sense of personal identity.
- Body identified as 3-space changes coherent to intent.
- Non-body (external universe) identified as those 3-space changes not responding to intent.
- Provides <u>*self*</u>, functioning in the <u>*world of non-self*</u>.

Energetics for superweapons

- Khrushchev demanded 100% ABM defense.
- Kapitsa replied to Khrushchev:

  "If a means of total neutralization of foreign missiles is to be found, it can only come from a group of new principles in physics, called energetics."

Basic Energetics

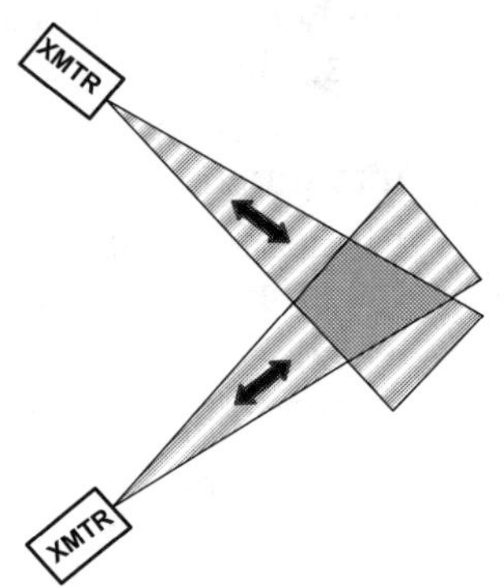

- Scalar interferometry
- Longitudinal EM waves
- Precursor engines
- Negative energy EMP
- Quantum potential
- Nonmaterial robots
- Carrier robots
- Mind control
- Mind conversion
- Building living systems including living minds

Selected energetics tests

- Scalar interferometry
- Kill of U.S.S. Thresher
- Deep underwater explosion
- Disease inductions at U.S. Embassy
- Tesla Shields (ABM defense)
- Cold Explosions
- Weather engineering
- Flameout of aircraft engines
- Kill of Arrow DC-8
- Kill of Challenger
- Kill of TWA-800
- Quantum potential test
- Mind tests — Button, Svoboda, Hess, etc.
- Robot tests, robot strategic war

Whittaker's inner electrodynamics* shows the precursor form in space (1)

- Combination of the two Whittaker papers "decomposes" all EM fields, potentials, and waves into a far more fundamental longitudinal EM biwave dynamics. The latter highly complex *continuous flow* structure is the engine or dynamic form of patterning of the field, potential, or wave and its dynamics *as it actually exists in mass-free space*.
- A longitudinal wave is an oscillating change of energy density in spacetime, hence an oscillating ST curvature. *Whittaker's structure is actually the form in which the spacetime precursor EM potential, wave, or field exists in space.*

* The 1904 Whittaker paper initiated superpotential theory. The 1903 paper has been essentially ignored.

Whittaker's inner electrodynamics* shows the precursor form in space (2)

- The fields, potentials, and waves taught in CEM/EE are the integrated charged matter fields, potentials, and waves existing only in charged matter when it is being acted upon by the Whittaker precursor engines.
- Integration of the precursor EM engine structure occurs only in charged mass.
- Combination of the two papers provides basis for a directly engineerable unified field theory (UFT), more fundamental than what is used in the West in its century-old electrodynamics and electrical engineering.
- Soviets began intensive development of this UFT immediately after WW II with a series of Manhattan-style projects.

Precursor scalar potential in space: longitudinal EM biwave sets

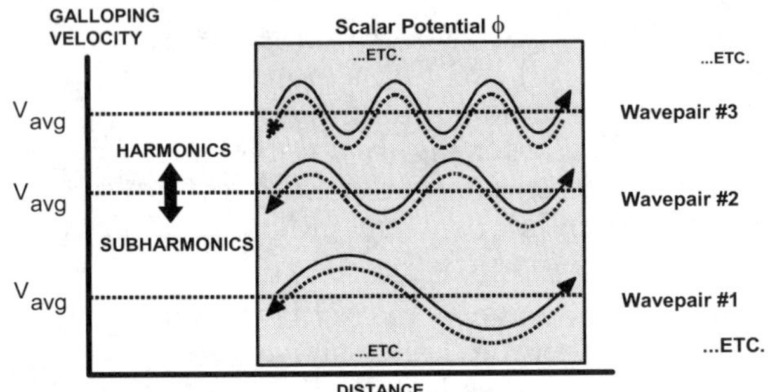

The Precursor Structure Is:

A harmonic set of longitudinal wavepairs. In each wavepair the two waves superpose spatially, but travel in opposite directions. The two are phase conjugates and time-reversed replicas of each other. Thus they comprise a coupled longitudinal wave and antiwave. The photons must be coupled into photon/antiphoton pairs (gravitons) by a strong application of the distortion correction theorem of nonlinear optics. Each wave in the biwave pair is a galloping wave. Each wavepair is a standing electrogravitational wave.

Note: Think of the oscillations as velocity modulations. The time-density carried by the waves is oscillating.

Quantum field theory

- Four types of photons (polarizations): x, y, z, t.
- Two (x, y) are transverse photons (familiar).
- Third is longitudinal (z) photon; energy oscillates to and fro along line of propagation. Bunches and thins energy density of space; is dynamic oscillation of curvature of spacetime.
- Thus is basis for engineerable integrated unified field theory.
- Fourth is scalar (t-polarized) photon. Energy oscillates to and fro along the time axis. Dilates and compresses time; hence is dynamic oscillation of curvature of spacetime.
- Scalar and longitudinal photons are individually nonobservable; but the combination of the two is observed as instantaneous scalar potential (common voltage).

Scalar interferometer*

- Energy effect in distant zone may be either heating or cooling, and slow (rising) or fast (explosive).
- Specific patterns of local ST curvature and dynamics (engines) may also be produced.
- Beams easily pass through the Earth or the ocean, whose internal fields and potentials are superhighways for internal longitudinal EM waves.

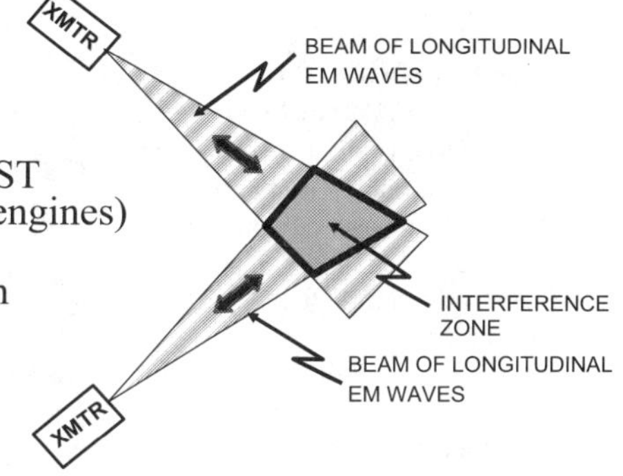

* Proof is given in M. W. Evans, P. K. Anastasovski, T. E. Bearden et al., "On Whittaker's Representation of the Electromagnetic Entity in Vacuo, Part V: The Production of Transverse Fields and Energy by Scalar Interferometry," J. New Energy, 4(3), Winter 1999, p. 76-78.

Khrushchev's First Demonstration Strike

- Kill of the U.S.S. Thresher. With her electronic controls jammed in the hash of a scalar interferometer whose IZ surrounds it underwater, the hull of the doomed submarine implodes when she reaches crush depth.
- The internal electronic jamming signals arise from the local spacetime everywhere within the sub (within the IZ).
- USS Skylark, companion surface ship nearby and in the "splatter" zone of the IZ, had multiple electronic systems fail, etc. totally anomalously.

Khrushchev's Second Demonstration Strike

Deep Underwater EM Explosion 11 Apr. 1963

(similar to this photo set of a deep underwater nuclear test explosion)

•Phenomena observed and reported by passing U.S. jetliner; later reported to FBI and to U.S. Coast Guard

•100 miles north of Puerto Rico, over Puerto Rican Trench (one of the deepest parts of the ocean)

Deep underwater nuclear burst, demonstrating same water phenomenology produced by Khrushchev's second demonstration on 11 Apr. 1963, the day after killing the U.S.S. Thresher.

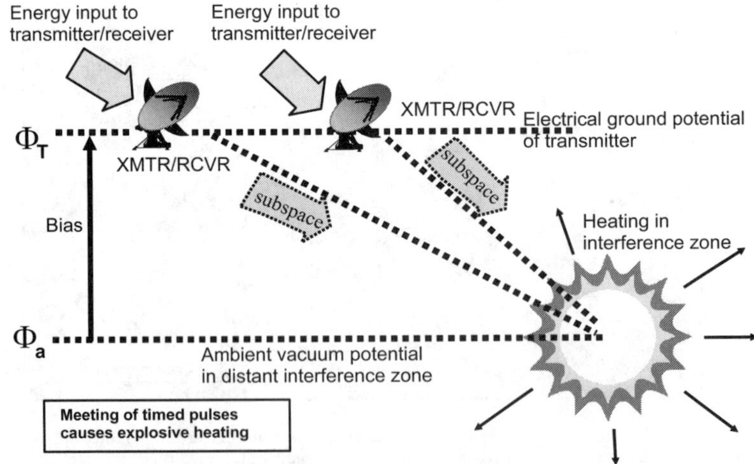

Scalar interferometry in the exothermic (heating) mode.

We may call the internal Whittaker electromagnetics a special kind of "active subspace", which it is. One can infold as many such "subspaces" inside normal EM fields, potentials, and waves as one desires. Hence one can engineer n-dimensional space with the new electrodynamics.

Scalar Interferometry In The Endothermic (Cooling) Mode

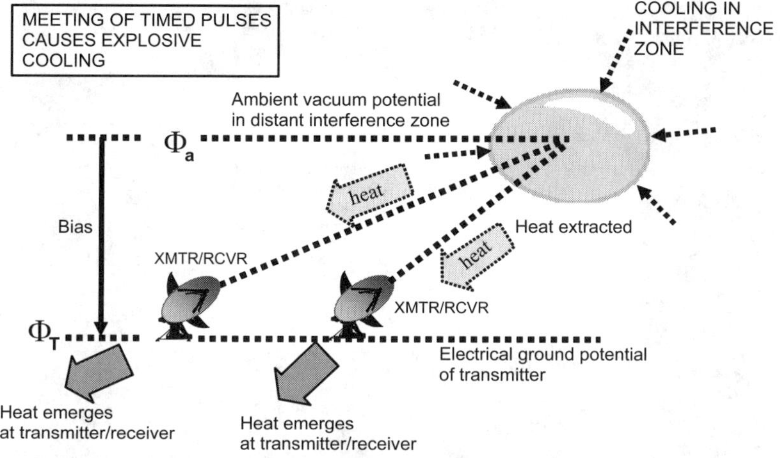

In the endothermic mode, tremendous heat is received back in the transmitters. Soviets further retransmit the heat via interferometers to a distant "exhaust" point. Bennett Island was used for such an exhaust point for decades, and our observation platforms saw a great many 150 miles-long jets of exhaust effects in narrow beams at shallow angles.

The Tesla Shield: Hemispherical Shell of Glowing Energy

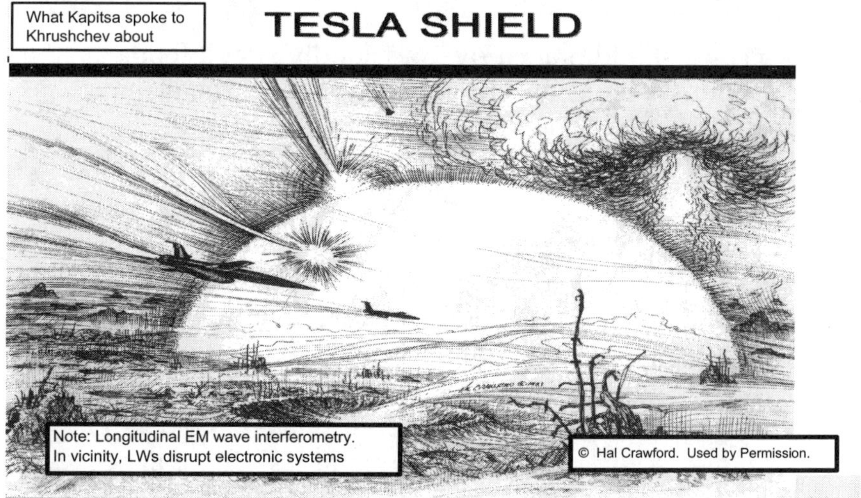

GIANT MISSILE SHIELD TEST SEEN BY PASSING SHIP'S CREW

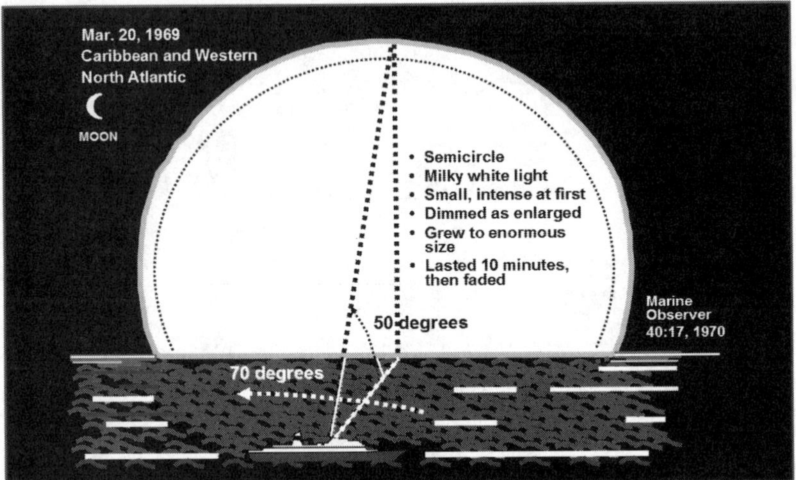

All the intense energy is in a thin hemispherical shell, between the outside edge and the inner dotted line. A missile or an aircraft passing through that intense energy zone is instantly destroyed, electronics dudded, warheads exploded, crew instantly dead, etc.

Special note on Tesla Shields (1)

- Tesla shield placed over a friendly site defends against incoming missiles, aircraft, warheads, etc. *If shield is made with "one-way out engines", defensive site can fire out and nothing can fire in*.
- Unrestricted Tesla shield placed over a hostile site, "bottles it up" against both incoming and outgoing vehicles, rockets, artillery shells, bombs, etc.
- If made *one-way in*, the bottling up allows outside firings into the targeted site, which itself cannot fire out. Useful against ICBM silos and similar targets.
- So it is very useful *offensively* as well as *defensively*. E.g., an entire naval task force at sea could be "bottled up" and neutralized.

Special note on Tesla Shields (2)

- Multiple internesting of up to three Tesla shields (shells) has been observed. Dramatically increases <u>*weapon kill*</u> probability against nuclear weapons incoming or outgoing and penetrating the shields.
- A giant single shield has been observed covering Atlanta, Georgia in the "bottle up" mode, but with minimal power so that no damage was done to aircraft penetrating the hemispherical shell in either direction.
- A shield is effective as a cover over hostile ICBM fields, to destroy their hostile ICBMs shortly after launch, while permitting some Russian ICBMs to penetrate and destroy the complex. Russia has had a dramatically effective launch phase ABM system for decades. So has the Yakuza, on-site in Russia.

Nick Downie's sighting from the ground of distant Soviet missile shield tests.

Missile Shield at Sary Shagan

THE SUNDAY TIMES, 17 AUGUST 1980

WARFARE

Witness to a super weapon?

HAS RUSSIA developed the ultimate weapon—the death ray, no less? In more technical terms, has the Soviet Union brought to testing stage what is called a "direct energy weapon?" And if not, what did British TV cameraman Nick Downie see in the sky above the peaks of the Hindu Kush in northernmost Afghanistan?

For years, there have been rumours that the Russians are working on a scheme to destroy incoming missiles not with rockets but with rays — beams directed like giant searchlights. Two systems are mentioned: high-energy laser beams or beams of sub-atomic particles, probably protons.

The most passionate prophet of doom has been the former head of US Air Force Intelligence, Major-General George Keegan. Having for some years claimed that the Soviet Union was mounting a major research effort in this field—2,000 physicists in 350 laboratories, he said—Keegan caused something of a stir a year or so ago by claiming that they were nearing the point of full-scale trials of a beam weapon.

Keegan's claims have been greeted with a scepticism muted only by three facts. The first is that his track-record as an intelligence officer was apparently first class. The second is that there is no dispute that in the United States as well as the Soviet Union, a good deal of work has been done on the military use of lasers and particle beams. The final point is that, if Keegan is right, the Soviet Union has achieved a very significant breakthrough indeed.

Both forms of the weapon would work in the same way. The beam—light in the case of lasers, particles in the other device—would need enough energy to bore through the skin of the missile. Then either the structural weakness round the tiny entry hole would cause the missile to break up; or the beam would proceed inwards to melt the missile's electronic innards. Because the beams would travel with the speed of light, they would reach their targets, to all intents, instantaneously and at enormous range. In effect, the Soviet Union could construct an anti-missile umbrella over its country.

That is the theory. The scepticism with which Keegan's views are greeted by many scientists and defence experts relates to a practical question: is there an energy source able to produce a beam of the required power at such range?

"To produce this sort of energy, you need an enormous power-pack," said one British expert. "It may be feasible in time to come but we just don't foresee a technology appearing on the horizon which is capable of it."

The debate has now reached a new intensity with assertions in the latest issues of the influential American magazine, Aviation Week and Space Technology, that the Russians are actually constructing an early prototype of a beam weapon—probably a charged-particle one—on their missile range at Sarychagan in deepest Kazakhstan near the Chinese and Afghan borders.

Unnamed figures in the US military are quoted at length. "If we were looking for a smoking pistol two years ago, we've got one now," one US beam-weapons expert is reported as saying. "Particle beams, as weapons, are real, and we can see the Russian machine taking shape from the overhead stuff" (i.e. reconnaissance satellites).

It has not escaped notice, however, that it is budget-allocation time back at the Pentagon—and what better way to persuade defence secretary Harold Brown to spend more on US beam research than to drum up a scare about Soviet progress? Aviation Week is indulging in "nothing more than an advertising campaign for the aerospace industry," says Kosta Tsipis, from the Massachusetts Institute of Technology. (Tsipis is co-author of an assessment of particle beam weapons which recently concluded their use would pose great, even insuperable, operational difficulties.)

Yet something odd is going on around Sarychagan. Nick Downie appears to have witnessed it. Downie—a former member of the SAS who has since established a considerable reputation as a war cameraman—was last September deep inside Afghanistan, at a mountain village called Shaltan. Across the valley lay the main range of the Hindu Kush and a hundred miles or so beyond that the Soviet frontier.

"It was twilight, and the villagers called me out excitedly to look at something they'd spotted in the sky," he says. "A bright reddish glow appeared on the horizon to the northwest behind a 10,000 foot ridge of the Hindu Kush. It was semi-circular, and resembled the moon, but was not a solid object. It was made up of light which gradually expanded until it occupied an arc of some 20 degrees. By the time it was high in the sky, it had lost much of its intensity. But what was very strange was that although the light gradually got fainter it retained a very sharp edge right until the end. Then, it faded out and disappeared."

Downie saw the same vision—it lasted 10-15 minutes—from another vantage point two weeks later. And he met an Afgh[an] who had seen it months befor[e].

What was it? Just possibly, [it] was a rare atmospheric sunse[t] phenomenon known as "purpl[e] lights." But a likelier explan[a]tion comes from Dr Ken Bignell, an atmospherics physicist a[t] Imperial College in London, wh[o] has questioned Downie. "I'[m] pretty sure it was a Sovi[et] rocket experiment."

"What Downie saw corresponds fairly exactly to wha[t] happens when a high-altitud[e] test is carried out. A smal[l] grenade containing, say, sodium, cadmium or lithium, is shot int[o] space and exploded. This spreads out in the upper atmosphere and according to factor[s] like temperature or ion densit[y] radiates a certain sort of ligh[t]. This is analysed with the aid of spectrometers on the ground."

Such tests give a valuabl[e] insight into the physics of th[e] upper atmosphere. They a[re] used in experiments on radi[o] communications, for exampl[e]. And they would be a standar[d] procedure in testing laser beam[s].

"What you would do i[s] launch a laser pulse from th[e] satellite and have it pointing a[t] some station on the ground an[d] then measure how much energ[y] arrives," said Dr Bignell. "How much of the energy of a lase[r] gets lost depends on the composition of the atmosphere."

Such tests, so high are the[y], would be visible for hundreds o[f] miles. The final clue, therefore, is the direction from which th[e] phenomenon came.

Downie took compass bea[r]ings. The glow was to the nort[h]-north west. He was looki[ng] straight towards the Soviet rang[e] at Sarychagan.

Gwynne Roberts

Nick Downie describes the strange lurid glow that flared silently over the Hindu Kush

London Sunday Times, 17 Aug 1980

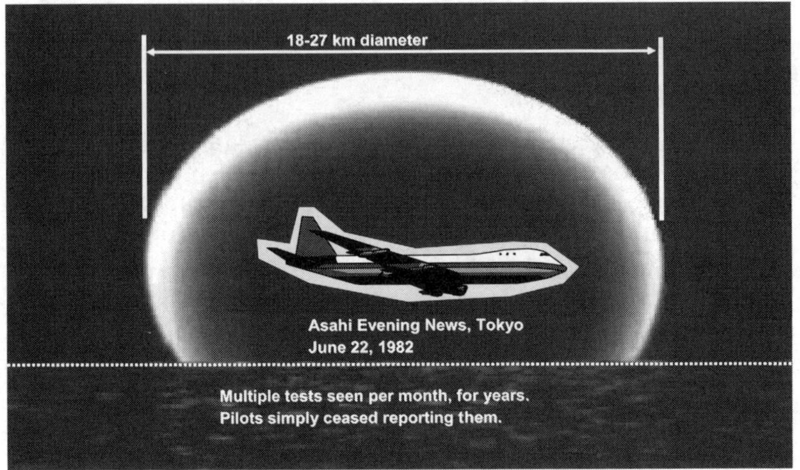

Typical shield test, northern route to Japan, seen by passing airliner. Many dozens of such tests have been seen and reported. Sightings were so common that pilots simply quit reporting them.

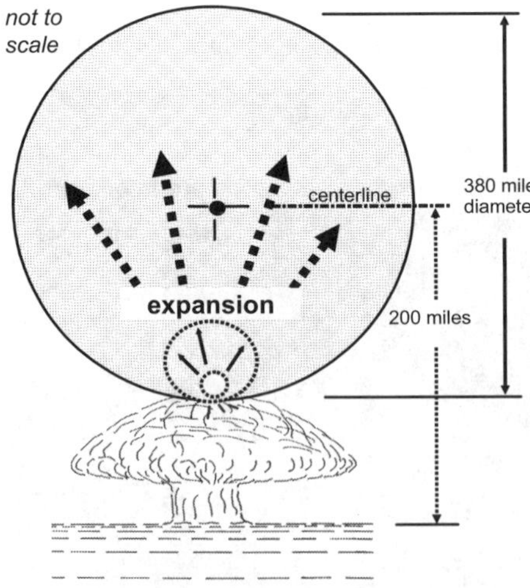

Sequence of Events

- Geyser of water rose rapidly from surface, mushroomed out
- Reached 60,000 feet altitude in about two minutes
- Brilliantly glowing small globe then appeared in top of cloud
- Globe increased greatly in size and altitude
- Reached diameter of 380 miles
- Center reached altitude of 200 miles

What They Were

- Superpotential interferometry using longitudinal EM waves
- Cold explosion just above surface (as antiship, antifleet)
- Smaller, higher density spherical shell (Tesla globe) for one-on-one shooting
- Larger Tesla globe for large area targets such as ICBMs coming in from midcourse
- Use in, out of atmosphere

Endothermic (cold) explosion off the Kurils 9 April 1984. Followed by a second test, an expanding globe of light, added in the exothermic mode to demonstrate rapid serial switching between weapon modes.

Weather engineering by scalar interferometry

INTERFERENCE GRID
© Tony Craddock 2004

- Steering jet streams
  - Make low pressure areas (endothermic)
  - Make high pressure areas (exothermic)
  - Move highs and lows, to entrain and steer the jet streams.
  - Displace cold air from Canada, warm air from Gulf, and where they meet provides sharp fronts and very violent weather.
- Sharp bends in jet stream add angular momentum.
  - Spawns tornadoes, powerful storms
  - Generates much atmospheric energy and lightning.
  - Ignites fires in those areas deliberately kept dry.

- Maneuvering and pulsing high and low pressure areas vertically above each other produces sudden downbursts.
- Many combinations and variations available and tested.

Vivid example of detailed structuring in weather engineering interferometry

Courtesy Scott Stevens, www.weatherwars.info

A few weather engineering signatures*

Sean MacLachlan

- Giant radial clouds, twin giant radials.
- Sharp "grid" of interference patterns across a giant section of the sky.
- High structure in a giant radial or twin giant radial.
- Moving giant radial or twin (crews sometimes just use a major highway, and move it along over that highway).
- Anomalous aerial booms, flashes, etc. when "adjusting a grid" or structure of a giant radial.

* A large number are documented in T. E. Bearden, "Soviet Weather Engineering Over North America", a 45-minute videotape. Huntsville, Alabama was a chosen "turning point" to turn the jetstreams up along the Southeast coast. Hence over the years we observed hundreds of cases and signatures and photographed dozens of them. Other present-day examples are on www.cheniere.org. Also, particularly see Scott Stevens, www.weatherwars.info.

Giant radial with internal structuring

Photo by Sean MacLachlan

Directional radial with structuring

Photo by Sean MacLachlan

Structuring in clouds

Photo by Marcia Stockton

More structuring in clouds

Photo by Marcia Stockton

Complex structuring in clouds

Photo by Marcia Stockton

Spectacular spiral cloud (1)

Photo by Tony Craddock

Spectacular spiral cloud (2)

Photo by Tony Craddock

Other geological uses of scalar interferometry (1)

- Deposit energy in fault zones, to generate piezoelectric pressure.
  - Quakes result when plates slip.
  - Slow increase gives great overpressure prior to slip; hence it produces a very strong quake such as 8.0 or even 9.5.
  - More rapid increase usually gives less overpressure prior to slip; hence medium quake.
  - Quite rapid increase gives tremors and small quakes to medium quakes at will, adjustable as to size.
- Deposit energy in rocks in a no-fault zone, and produce anomalous quakes. Can vary magnitude by adjusting rate of energy deposit, as stated.

Other geological uses of scalar interferometry (2)

PLUME 03-08-05
FROM MOUNT ST HELEN

- Initiate volcano eruptions by deposit of energy.
- Use sea-bottom quakes to generate tsunamis.
- When this first written (July 27, 2003), many volcanoes around the world were suddenly active. Still are.
  - Too many to be totally "natural".
  - Were and are Yakuza practice and *preparations.*
- Highly destructive 26 Dec 2004 tsunamis induced in Asia were a Yakuza opening test for the Operations Phase of asymmetric war just entered. They induced the 9.3 quake in the ocean floor causing the tsunamis.

How much does weather and climate affect us?

- *"40% of the $10 trillion U.S. economy is affected by weather and climate changes."*
  Conrad C. Lautenbacher, U.S. Undersecretary for Commerce, Wall Street Journal, July 28, 2003.
- So in addition to what nature's changes already do, those weather and climate changes being augmented or implemented by hostile weather engineering are costing us a bundle.
- This is a special kind of war, waged in "peacetime", and frightfully damaging economically.
- Weapons use confirmed in 1997 by SecDef Cohen.
- It's just part of the ongoing "bleed the U.S. dragon".

Other uses of scalar interferometry (2)

- Deposit negative EM energy in fault zone where pressure is rising naturally or by weapons.
  - Extracts energy from the zone.
  - Reduces pressure, avoiding pending quake.

- Deposit EMP in control center of power plant,* such as a large nuclear power plant. Controls are dudded, runaway reactor core risks meltdown.
- Generate explosions in refineries, oil fields, either of positive energy or of negative energy.
- Use in bioenergetics and psychoenergetics attacks as well (produce *engines* in the interference zone).

* Negative energy EMP destroys all solid state electronics instantly. A strong negative energy EMP kills generators, current devices, and all living systems. Since nuclear binding energy is negative, such EMP arising in spacetime in the interference zone can and does directly affect and perturb nuclear material in the core. or in stored control rods. If specific "engines" are also incorporated, the core runaway can probably be induced directly.

Other uses of scalar interferometry (3)

- Scalar interferometer beams and signals can be "imbedded" in two or more normal but powerful EM signals from "normal" transmitters, and used wherever the signals meet or overlap.
- Can also operate totally "inside" a normal static potential, EM field, etc.
- Thus scalar interferometry was "buried" and disguised inside the Soviet Woodpecker signals.
- U.S. CEM/EE measurements of the overall Woodpecker signals showed nothing of it.
- Hence our nation slept on, sublimely ignorant in its belief in its archaic and flawed CEM/EE.

Brewer Briefs President Reagan on Soviet Scalar EM Weapons Oct 84
Then Briefs National Security Council

Typical ABM Defenses
(widely deployed by mid-70s')

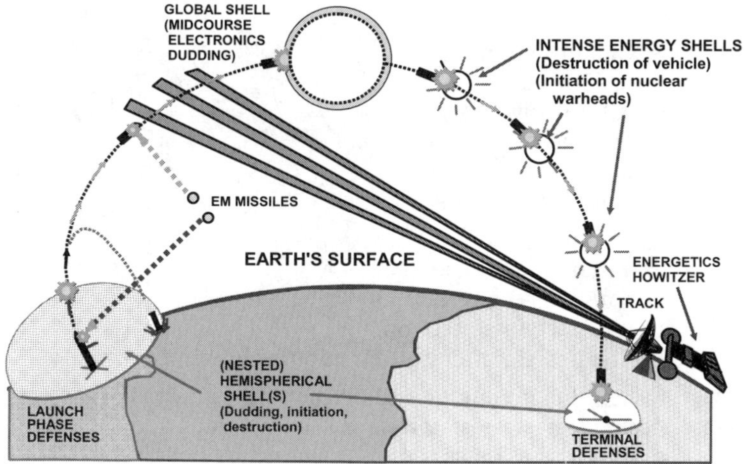

Vacuum engine (precursor engine) concept (1)

- Wheeler* said:
  "*Space acts on matter, telling it how to move. In turn, matter reacts back on space, telling it how to curve.*"
- Wheeler's principle can be expanded for a whole set of spacetime curvatures and their dynamics, leading to the <u>*spacetime curvature engine*</u> concept.
- This also affects virtual particle flux of vacuum, so it creates an exactly corresponding vacuum engine.
- There is no observable mass in space, so these are "massless engines"; i.e., they are precursors to the force engines that result in matter when the precursor engines interact with it.

* Wheeler's principle, as stated in W. Misner, K.S. Thorne, and J.A. Wheeler, <u>Gravitation</u>, W.H. Freeman and Co., San Francisco, 1973, p. 5.

Vacuum engine (precursor engine) concept (2)

- A vacuum engine or precursor engine is a structural form in spacetime, consisting of curvatures and curvature dynamics.
- The "pattern" and dynamics for many actions, to be induced in mass when acted upon by the precursor engine, is captured by the geometry.

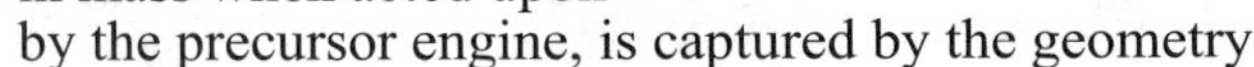

- When this precursor engine then interacts with mass, it creates an exactly correlated set of forces and their dynamics, acting in that mass.

Vacuum engine (precursor engine) concept (3)

- <u>*Said another way:*</u> Levels of dynamic energy structures and levels of dynamic time structures mold spacetime geometry itself.
- This precursor "pattern" and its dynamics is called a *template* or *engine*.
- The precursor template interacts with, and structures, the virtual particle flux of vacuum.
- This forms an exactly corresponding <u>*vacuum*</u> engine (still a special precursor template).
- Spatiotemporal structurings and dynamics of the engine structuring and dynamics of the VPF act upon mass at all levels.

Engines (vacuum engines, spacetime curvature engines)

- Normal EM fields, potentials, and waves decompose into sets of longitudinal EM waves (LWs) and their dynamics.*
- Since these LWs and their dynamics are dynamics in energy density of spacetime (ST), they are *dynamic ST curvatures*.
- They enfold general relativity inside extended electrodynamics.
- By manipulating and changing these LWs, one manipulates and changes spacetime curvatures and their dynamics (and virtual particle flux of vacuum and its dynamics as well).
- It is a Unified Field engineering where one controls and engineers a set of ST curvatures by use of higher group symmetry electrodynamics.

* E.T. Whittaker, Math. Ann., 57, 333-355 (1903); — Proc. Lond. Math. Soc., Series 2, 1, 1367-372 (1904).

Signal vs. precursor vacuum engine: bioenergetics example

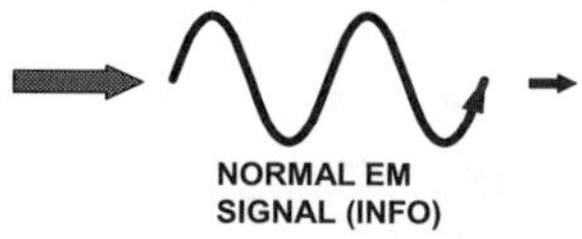

- Information received
- Any overt physical action must be taken by cell itself
- Cell must furnish the energy for any action it takes
- Passes from outside to in

A. *Cell must do the action itself; vacuum energy exchange is passive. Energy or fuel for doing the action must be added to cell externally.*

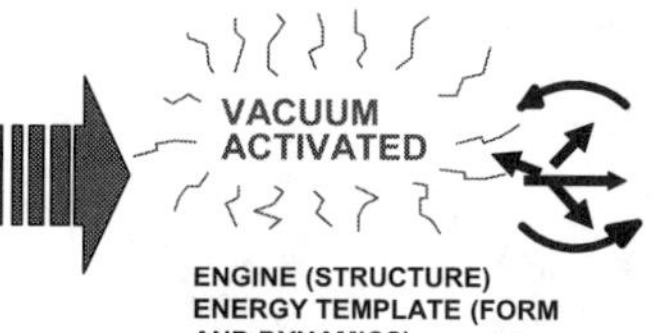

- Information itself acts
- Local spacetime curved, with internal subcurvatures
- Produces template as local spacetime (ST) curvatures
- Produces forces in the interacting matter
- Acts on and changes all parts of cell, including its genetics

B. *Cell is acted upon; vacuum energy exchange does the specific action. No energy or fuel for doing the action need be added to cell externally.*

Unified field theory requires modeling and use of *supersystem*

Three supersystem components interact with each other:

- System and its force dynamics
- Local active vacuum and its precursor energy dynamics (and engines).
- Local curvatures of spacetime and their precursor dynamics (and engines).

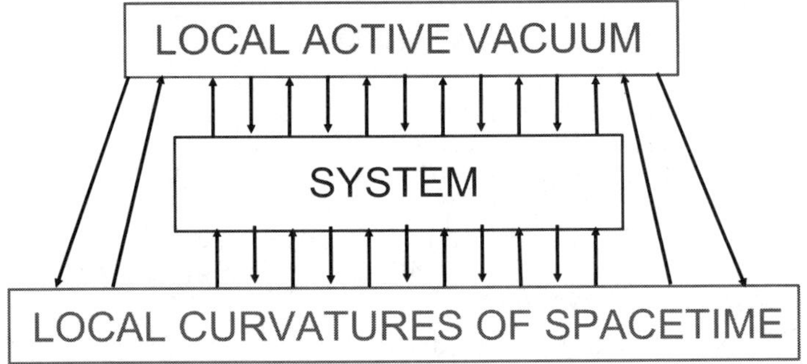

Classical EM kills the supersystem, prevents supersystem engineering

- Assumes flat local spacetime
- Assumes inert vacuum
- Assumes no internal precursor energy engines.

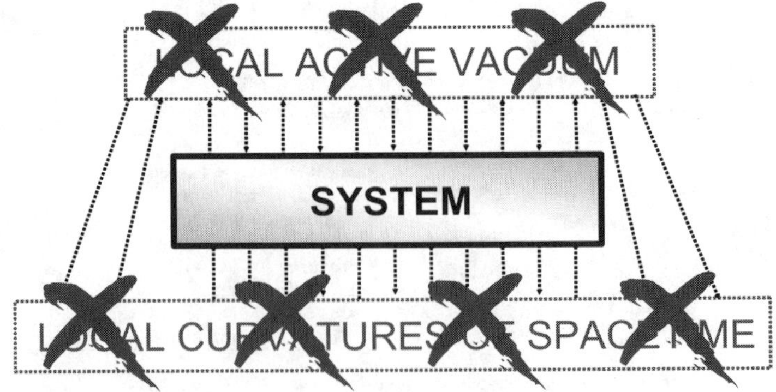

Modeling the supersystem

- Higher group symmetry electrodynamics must be used.
  - Quaternions as a minimum.
  - O(3) EM developed by Evans.
  - SU(2)×SU(2) EM favored by Barrett.
  - Clifford algebra EM shown by others.
- The algebra selected must be capable of modeling the <u>*supersystem*</u> interactions (system, active vacuum, curvatures of spacetime) and their structural dynamics.
- It becomes an engineerable grand unified field theory. Evans' new GUFT appears particularly noteworthy.
- Russian energetics is precisely such a model, obtained by correcting the known serious errors in CEM/EE.

Inertial field device of Toronto inventor Sid Hurwich*

- Hurwich gave Israelis a secret device in 1969.
- Demonstrated to Toronto police in 1969.
  - Physically "jammed" a pistol lying on a table, so it could not be moved or fired.
  - After a half-hour in the room for the experiment, *clocks and watches had not changed their time from time of entry*.

- Hurwich was given "Protector of the State of Israel" award, June 1976.

- Invention used at Entebbe, July 3, 1976.
- Used again against Iraqi nuclear reactors.**

---

* David Jones, "Israel's Secret Weapon," <u>Vancouver Sun Times</u>, Weekend Magazine, Dec. 17, 1977, p. 17.

** Private U.S. source who observed aspects of it.

Novel capability from "freezing time" (1)

- The flow of time slows in a rotated frame.
- Rotating frame by 90 degrees "freezes" time. Rotated object appears as just "light" in the lab frame.
- To rotated observer, lab itself appears as "just light"; can walk through it.
- Hurwich only affected the inertia of some objects in his early device, but effectively.
- In one experiment Golden affected clocks and watches, of all types, for four days.

Novel capability from "freezing time" (2)

- Operating in one's own *locally* rotated orthogonal frame, one can "step around" a 3-space lab wall.
  - In *lab* frame one appears to pass through a wall.
  - In one's own frame, the wall is "of light" and one merely steps through that light.
  - Rotated person sometimes referred to as "cloaked".
  - Bullets pass harmlessly through a cloaked person.
- At least two nations have such capability, as does a not-so-nice group using it for assassinations.
- We had a friendly demonstration and two hostile demonstrations (surviving the latter by blind luck).

Some demonstrated capabilities 

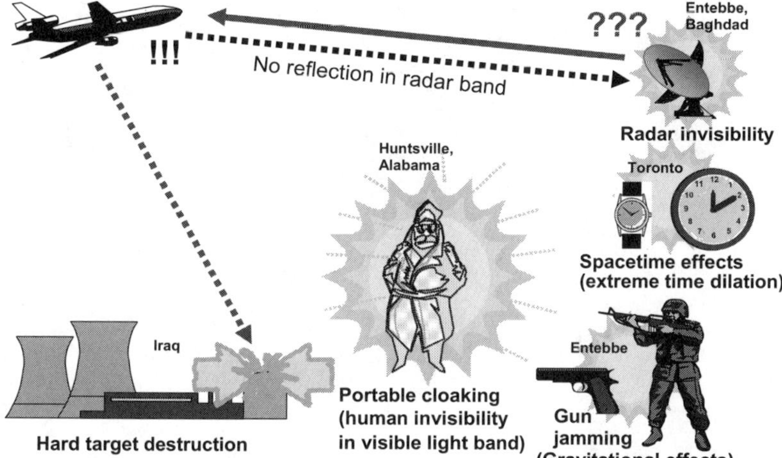

(SLIDE BRIEFING CONTINUED)

PART C: Move Toward First Strategic Energetics Attack

MOVE TOWARD FIRST STRATEGIC ENERGETICS ATTACK

Move toward first strategic energetics attack

• • • • • • • • • • • • • • • • • •

Brezhnev's timetable and slowly escalating incidents

Section contents

- **Schedule**
- **Typical test strikes**
- **Characterization and capabilities**
- **Unexpected retaliation**
- **Israeli retaliation**
- **Scheduled strike aborted**

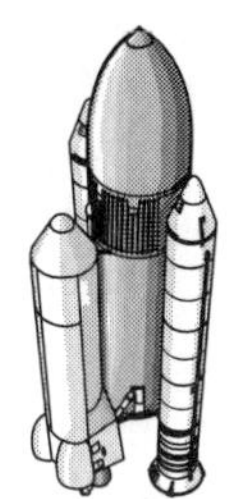

Khrushchev's Jan. 1960 statement to the Presidium

- *"We have a new weapon, just within the portfolio of our scientists, so powerful that, if unrestrainedly used, it could wipe out all life on earth."*
- *"It is a 'fantastic' weapon."*
- Ever the ebullient peasant, Khrushchev sought to very quickly change the balance of power.
- Impatiently moved before his scalar interferometers were in operation.

Generation of the Cuban missile crisis

- Khrushchev inserted long range missiles and nuclear warheads into Cuba in latter 1962, to change the balance of power quickly.
- His strategic scalar interferometers were not yet deployed and ready. His nukes and ICBMs were also in bad shape.
  - Kennedy called his hand, and he was caught bluffing.
  - After blustering to get U.S. agreement not to invade Cuba, he was forced to back down, suffering severe humiliation before the world and the Communist Party.
- When his strategic interferometers came on line, he unleashed two demonstrations to recover face.
  - Killed the U.S.S. Thresher on station underwater, off the East Coast of the U.S.
  - Produced a giant underwater EM energy burst the next day, 100 miles from Puerto Rico.

Brezhnev's schedule:

- *"We are achieving with détente what our predecessors have been unable to achieve with the mailed fist...*
  ...by 1985 our power will be so irresistible that we can do what we wish anywhere on the globe."
- A decade later, at a secret meeting of European Communist Parties in 1972, Brezhnev laid out the 1985 goal and time table.
- In 1985 the Soviets would be ready to move to dominate the earth.

1985 Soviet preparations

- For May Day 1985, a huge Soviet exercise placed 27 "power taps" in the Earth, powering up scalar EM weapons arsenal.
- Each tap estimated to power up to 6 huge weapons.
- In 1985, initiated earthquakes, extensive weather engineering, killed aircraft and missiles, tuned up metal softening signals, initiated a series of highly anomalous storms and electrical activity in the atmosphere. Test shots offset from shuttle launch killed Arrow DC-8 at Gander, Newfoundland.
- In 1986, stepped up kill of aircraft, missiles, rockets, etc.
- Israel had struck Russians hard in 1984, in one strike destroying one third of all missiles in Soviet Northern fleet.
- Israel also checked the Soviet strategic attack plans shaping toward mid-to-latter 1986.

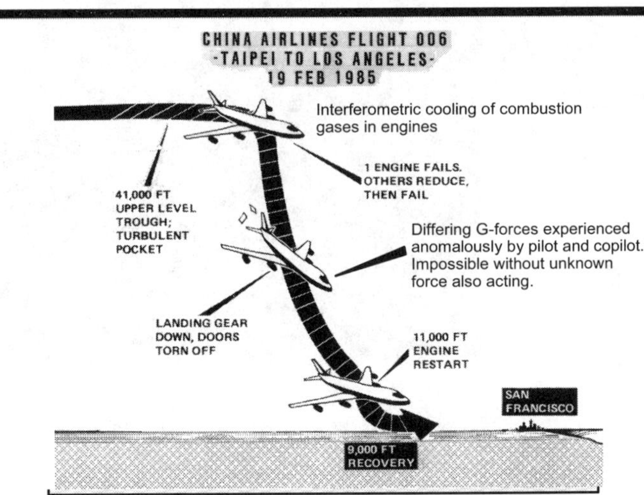

Cooling and extinguishing combustion in the engines of a jetliner 19 Feb. 1985. Use of scalar interferometry in endothermic mode.

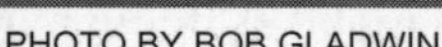

Scalar interferometer operator fires offset EM missile from Russia, through subspace interferometry.

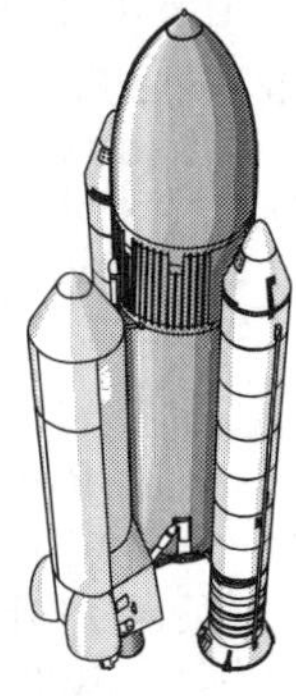

Actual Soviet EM missile strike offset from U.S. shuttle launch at Cape Canaveral on 26 Nov. 1985. This was the smoking gun.

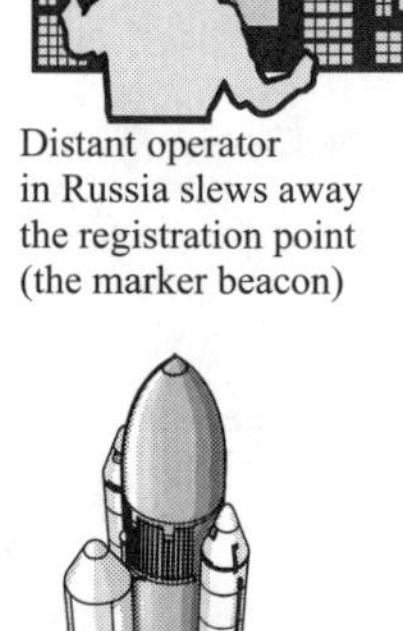

Distant operator in Russia slews away the registration point (the marker beacon)

26 Nov 1985

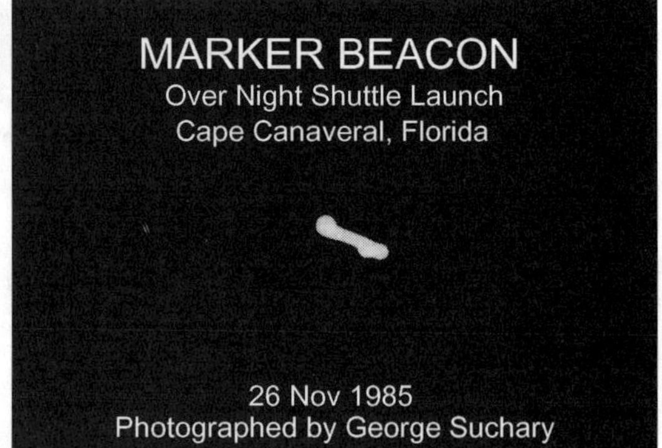

Marker beacon above U.S. shuttle launch at Cape Canaveral, 26 Nov. 1985. Russian crews were using shuttle launches as practice ICBM launch targets. This is a standard artillery "shift from known registration point" targeting.

Two weeks later, the same weapon killed the arrow DC-8 at its takeoff from Gander, Newfoundland on Dec. 12, 1985, and demonstrated meeting schedule.

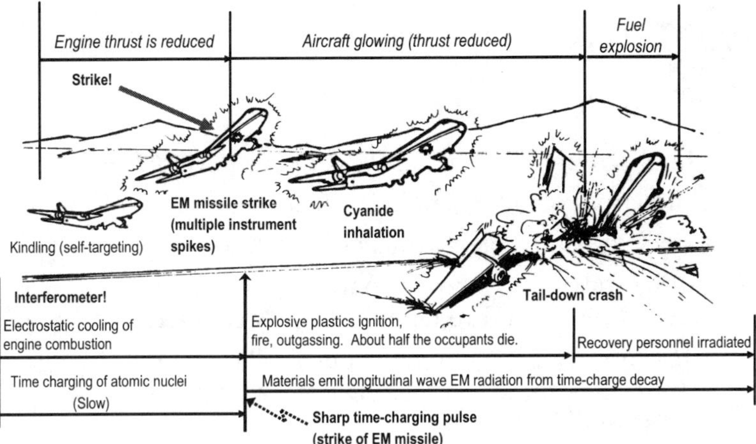

Kill of the Arrow DC-8 at Gander, Newfoundland Dec. 12, 1985. Dual use of scalar interferometry in endothermic mode (cool combustion in engines to reduce lift) and exothermic mode (strike with EM missile, as at Cape Canaveral 2 weeks earlier by same weapon).

Kill of Arrow DC-8

- Demonstrated meeting Brezhnev's schedule. Killed 248 U.S. soldiers, 8 crew personnel.
- Interferometer weapon tested in both modes:
  - Endothermic mode (cooling combustion of engines).
  - Exothermic mode (strike of EM missile).
- Successful as high level probe.
  - NSC was caught thinking it was Arab bomb, thus having to hide its illegal Iran-Contra activity.
  - U.S., Canadian scientific community totally asleep.
  - Political community interested in "burying" the incident quickly, as "icing accident".

- U.S. pressured Canadian government, ignored eyewitness of strike (not allowed to testify to investigation board), sent in a General Officer to bulldoze over the site quickly, withheld autopsy reports from Board.
- Ignored decisive later signatures such as dozens of handling personnel sickened by longitudinal EM wave radiation from the struck parts of the aircraft.

Kill of space shuttle Challenger Jan. 28, 1986

- Metal-softening signal on Woodpecker signal grid.
- Cold stress; Soviets engineered cold wave into Florida.
- Booster seal problem (thus aggravated).
- Soviet trawler spyships departed at high speed 4 hours before launch. Unprecedented.
- Birds not flying (unprecedented). Shows high frequencies for sharp localization, which pains bird's small brains.
- That night, KGB headquarters gave a party in Moscow to celebrate "perfect success of their active measures against the Challenger".*

* Per General Daniel O. Graham, Director of Project High Frontier and former head of Army intelligence. General Graham died in 1996.

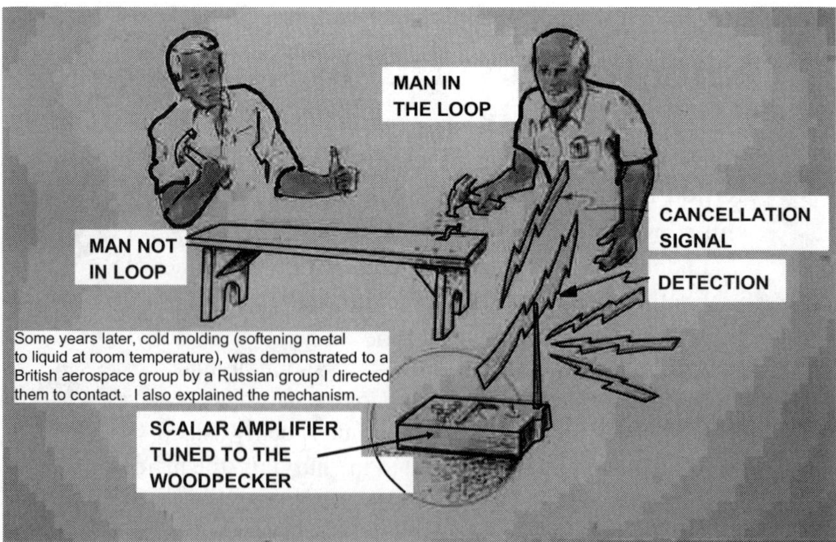

Proof of metal-softening signal: In Tennessee, Frank Golden softens nails with amplified Woodpecker signal in his area when Challenger destroyed in Florida.

U.S. airstrike on Libya, April 1986

Encountered test of a lab prototype quantum potential weapon.

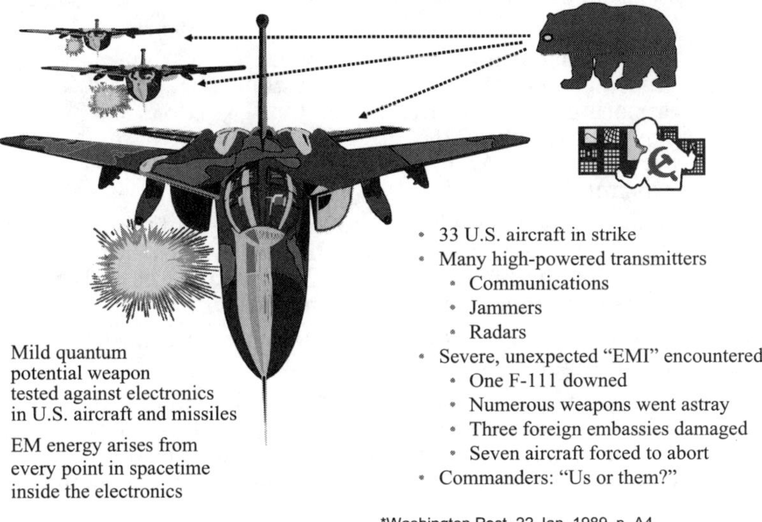

*Washington Post, 22 Jan. 1989, p. A4

Chernobyl disaster - Apr 25-26, 1986

- Great nuclear disaster at Chernobyl
- One of the nuclear reactors suffered a disastrous venting of its containment building
- Spewed out radioactive material, which spread extensively.
- Casualties, etc.
- Was pre-emptive strike by a private U.S. group.

What caused Chernobyl's nuclear reactor disaster

GHOST TOWN NEAR CHERNOBYL TODAY

- Russian scalar interferometers were kindling energy into San Andreas fault, preparing to give California its long-expected Richter 9.0+ quakes in San Francisco, Los Angeles, etc. "Big One" was imminent.
- Detected by a certain private U.S. group able to use a novel countermeasure. Considered striking the distant transmitters to prevent perhaps 100,000 to 200,000 U.S. casualties and grave destruction. Risk assessment: Medium.
- Group struck one of the transmitters (near Chernobyl), destroying it.
- Safety circuits on the transmitter held, bled off energy for nearly 24 hours, then failed. Flashover of remaining energy into earth struck Chernobyl reactor and caused its abrupt increase in excess radiation, venting the containment.

Chernobyl disaster 25-26 April 1986

An intense countermeasures pulse EMP from the U.S., backtracking on the interferometer signal wave, struck the transmitter(s), destroying one.

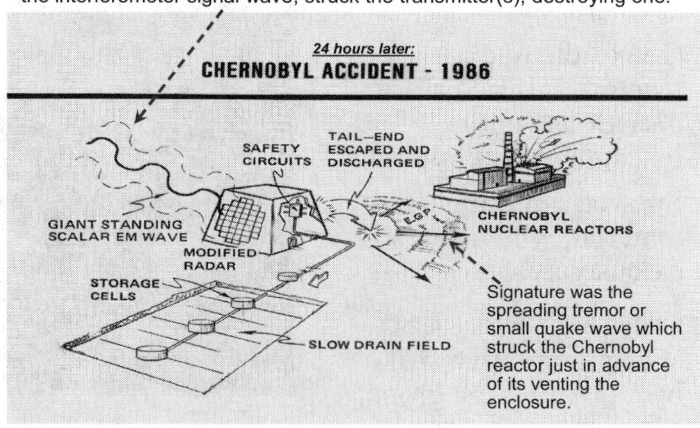

Israel forced Soviet 1986 attack plans to be aborted

- On Brezhnev's schedule, Soviets were prepared by the end of 1985 to attack and destroy the West in 1986.
    - Kill of Arrow DC-8 in Dec. 1985 demonstrated meeting the schedule.
    - 1984, '85, '86 very active in scalar weapons tests.
    - Middle East was to play an important role.
- Israel struck hard at Soviet forces in 1984; one strike alone destroyed some one-third of all the missiles in the Soviet Northern Fleet. This delayed Soviet attack plans.
- With Israeli QP superweapons facing it, Soviet Union (KGB) aborted strike plans to avoid being destroyed.*

* Development, operation, and control of the Russian energetics weapons have always been by the KGB, not by the regular Russian armed forces. Today the KGB is known as the FSB (Federalnaya Sluzhba Bezopasnosti), or in English, Federal Security Service.

Continued preparations for new attack posture

Chessplayers think many moves ahead; the game continues.

- **Tesla shields and typical tests.**
- **Earthquakes and Cohen's 1997 statement.**
- **Quantum potential and weapons capabilities.**
- **EM biological warfare.**
- **Dudding nuclear weapons.**
- **Aum Shinrikyo and Yakuza involvement.**
- **Typical test kill (TWA-800).**

Blunt warning from the Soviet Union

- *"If the United States deploys a shield in space, the Soviet Union will have several options, none of what Washington would wish... The Soviet Union will very quickly find a response of which the United States has no inkling as yet."*

Soviet Chief of Staff Sergei Akhromeyev, in the Washington Times of Aug. 26, 1986, warning the United States that the Soviet Union could have an unpleasant surprise response if the U.S. deploys its Strategic Defense Initiative (SDI) in space.

Comment: At the time, the Soviets had inserted nuclear weapons of 20 KT to 40 KT into most major U.S. cities, along with Spetznaz teams to detonate them on order. Those weapons and their teams are still here. See Lunev's book for some of the methods used to get the weapons into the U.S. In addition, at the time, the Soviet giant strategic scalar interferometers had been deployed for 23 years.

Shield test over Atlanta, Georgia 12 Nov. 1986

"Bottling up" shield practice over Atlanta 12 Nov. 1986.
With a real shield, aircraft or missiles flying into it
(from inside or from outside) explode and are destroyed.

Put up or shut up: Counting rifles with earthquakes

(One of the aftermaths of the 1986 Chernobyl counterstrike)

Opening bet (induced quakes)

- Quebec 25 Nov 88
- Pasadena 3 Dec 88

Raise you

- Armenia 7 Dec 88

Call - what do you have?

- Malibu 8 Jan 89

Four aces

- Soviet Central Asia 23 Jan 89

Clear table (stress relief)

- Bakersfield CA et al. (various)

SecDef Cohen confirms scalar interferometry weapons activity

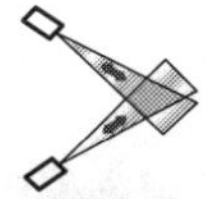

"Others are engaging even in an eco-type of terrorism whereby they can alter the climate, set off earthquakes, volcanoes remotely through the use of electromagnetic waves... So there are plenty of ingenious minds out there that are at work finding ways in which they can wreak terror upon other nations. It's real, and that's the reason why we have to intensify our efforts, and that's why this is so important."

*Quoted from DoD News Briefing, Secretary of Defense William S. Cohen, Q&A at the Conference on Terrorism, Weapons of Mass Destruction, and U.S. Strategy, University of Georgia, Athens, Apr. 28, 1997.

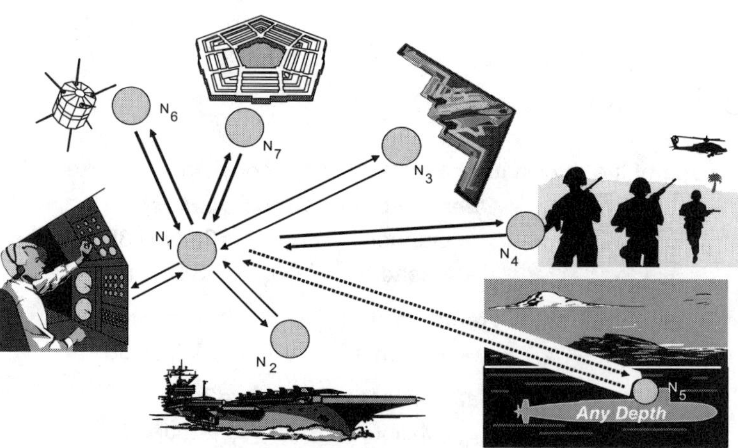

Strikes through multiply-connected spacetime. One joule of energy inserted into the lab connector, also simultaneously appears at each and every distant connected target. Tremendous "force amplifier" in military terms; can strike a thousand or million targets at once, using the energy one would only use for a single target in conventional strike.

Quantum potential instantly attacks widely separated targets. Embedded specific engines perform specific functions such as dudding nuclear warheads, propulsion, powerplants in 10 minutes.

HVT quantum potential (1)

- Bohm's quantum potential:

$$V_{quantum} = (\frac{\mathrm{h}^2}{2m})\frac{\tilde{\mathrm{N}}^2\sqrt{r}}{\sqrt{r}}$$

- *"A quantum particle moves as if it were subject, in addition to external potentials, to a potential which is a function of its own probability distribution."*

Assumptions of Bohm's quantum potential

- Particle and wave function are real and separate.
- Wave function obeys Schrödinger's equation.
- Particle obeys classical mechanics.
- Particle couples to wave function through a quantum potential.

$$(r = \mathrm{Y}\,\bar{\mathrm{Y}})$$

Quantum Potential Characteristics

- No point source, not radiated.
- For QP between two particles:
  - Interaction does not vanish as spatial separation becomes very large.
  - Instantaneous connection through multiply-connected spacetime.
- Depends on quantum state of system as a whole.
- System parts can be greatly separated.
- Most powerful weaponry on earth. Five nations* have it deployed, sixth is closing fast.

* Russia (KGB), Brazil, China, Israel, and one other.

At the infinite velocity of the quantum potential:

Each Is Every Other

Any point in the universe is sitting in the palm of one's hand

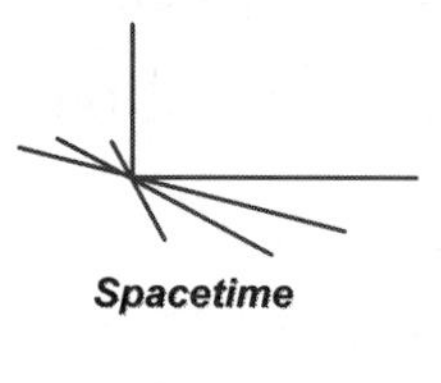

Spacetime

Virtual Particle Flux

Potential

Mind and Mind Levels

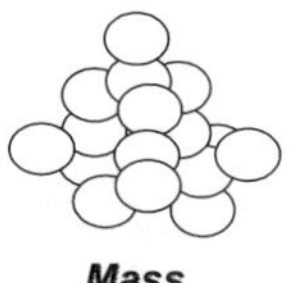

Mass

Planet

Living Organisms

Usefulness of quantum potential (QP)

- Physical reality has vast spectrum of QPs.
  - A QP connects every member of the human species.
  - All atoms of a specific kind have a weak QP.
  - All systems of a specific kind have a weak QP.

- All human minds have a QP, as do all human bodies.
- Soviets can form a QP among desired targets in a given area, and strengthen that QP in that area at will.
- Effect produced by QP weapon in one lab target is simultaneously produced in all participants in the QP.

Usefulness of quantum potential (QP) (2)

- QP weapons can attack a vast array of widely separated targets with one shot.
- A "mix" QP can be formed by "mixing" multiple component QPs.
- QP can contain a specific precursor engine, or a group of them.
- QP weapons are most powerful weapons on earth.
  - Could indeed destroy the entire human species with a single shot, killing every human everywhere.
  - Or could destroy every AIDS virus and every smallpox pathogen with a single shot.
- Can dud every nuclear weapon, propulsion system, and power plant in about 10 minutes.

Radionics: Elementary QP use

T. Galen Hieronymus

- When a body interacts with an external object by photon interaction, the body and that object are linked by a very weak QP that is established.
- A photo of a person is directly QP-linked to him.
- Hieronymus patented use of effect in the 1940s.
  - Killed insects in distant field by treating photo of the field with insecticide in his radionics machine producing a weak QP.
  - Cut out checkerboard pattern; killed bugs only in sections that were intact in segmented photo. Bugs migrated to the holes.
- Ukako corporation killed bugs for farmers for a decade as a useful commercial service, in multiple states.
- Was outlawed when states realized could also kill humans.
- <u>*Bin Laden and other high terrorists, e.g., could be killed instantly right where they are, merely from their photographs.*</u>

Quantum potential is a fantastic energy-amplifying weapon

- Use one primary "station" in operations station.
- Connect primary station to a large number of objects or systems in many remote places.
- Insert a one megawatt EM pulse into primary.
- Each and every distant object or system simultaneously has that same 1 MW of energy appear in it instantaneously, because of multiply connected spacetime (effective "superposition" of all participating targets).
- If attack a million targets, it is the same as attacking the "primary" and simultaneously all others with the same single attack. This is a vast "force amplifier" effect.
- 1 MW input in one local target gives 1,000 gigawatts total "output" in distant targets. Nothing else compares.

EM biological warfare

- Create Kaznacheyev's "disease engine" by pumping diseased cell in the time domain, then phase conjugating once again in time domain.
- Infold the disease engine in a carrier as a Whittaker-Devyatkov structure.

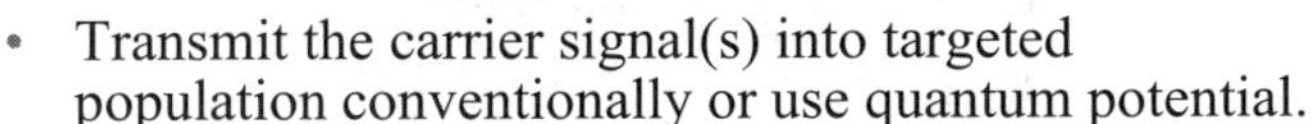

- Transmit the carrier signal(s) into targeted population conventionally or use quantum potential.
- Kindling of the disease occurs in the target bodies, inducing the disease. Gulf War disease induced this way.
- Note: Quantum potential (QP) is an ideal carrier; can strike entire populace at once, with disease engine(s) desired.
- QP with one robot carrier and its engines can instantly strike all targets on earth with reproduced robot/engine copies.

Countering EM BW: When signal is available

- QP defense is like dueling scorpions: He who strikes first to the vitals, wins.

- Receive, phase conjugate by time-pumping.
- Embed amplified phase conjugated engines in signals for transmission.
- Transmit amplified phase conjugated engines.
- Variation: put the antiengines inside the hostile transmission signal itself and fire.
- Places it back into the hostile transmitter(s).
- Hostile transmitter then transmits both the engine and antiengine, nullifying its own effect.

Possible advanced EM BW

- Covert EM BW using disease engines.
  - Materials activated (dimensioned).
  - Water structured (dimensioned).
  - Radio, TV emissions structured.
  - Electrical power grid signals structured.
  - Use of quantum potentials with engines.
- Mindsnapper (negative energy EMP).
  - Tested twice by Russians in Afghanistan.
  - Instant and total death: Separates all mind-aspects from every living system in struck area.
- Enhance or nullify vaccines and antibiotics.
- Enhance or nullify BW strikes in an area.

Indicators in EM BW

- Gurvich: Mitogenetic radiation
- Kaznacheyev: Cytopathogenic mirror effect
- Devyatkov: Information content of the field
- Whittaker: Infolded LW electrodynamics
- Lisitsyn: Genetic brain code broken
- Popp: Master cellular control system
- Sakharov et al.: New approaches to general relativity
- Becker: Cellular regenerative system
- Sachs: Unified field theory
- Evans: O(3) electrodynamics
Grand unified field theory (engineerable)
- Bearden: Engines using infolded LWs

QP weapons summary (1)

- Induce health changes and diseases in entire focused national populace at once. Has been tested in U.S.
- Expose immune systems to "shadow" (virtual) disease engines of many diseases at once. Immune system "spreads", thus totally vulnerable to any BW attack or opportunistic infections (Gulf War Disease was such a test). Tested in U.S.
- Dud all nuclear *weapons*, *propulsion systems*, *power plants* on earth or in an area in 10 minutes.

- As vacuum engine carrier, induce any desired effect simultaneously in millions of selected targets.
- Can attack physical systems, minds, bodies, electronics, nuclear, etc. by its internal engine(s).
- Most powerful weapon on earth; 5 nations now have it: Russia, Brazil, China, Israel, and one other friendly nation. A sixth nation is closing on it rapidly.

QP weapons can nullify:

- Nuclear dead-man fuzing weapons.
- ICBM nuclear warheads and electronics.
- Strategic bomber nuclear weapons and electronics.
- Nuclear submarine nuclear weapons and electronics.
- Nuclear submarine propulsion systems.
- Electrical power systems.

- Electronics and communications systems.
- Specific chemical agent contamination, biological agent contamination, or vaccines.
- Human immune systems.
- Satellites, ships, planes, subs, rockets, radios.

Simultaneous dudding nuclear weapons worldwide, and nuclear reactors and nuclear propulsion systems.

- Via quantum potential joining U235 and Plutonium in warheads.
- Includes weapons airborne, underwater, in storage.
- Transmutations of the nuclides themselves by inserted engines.
- Part of the initial barrage to be launched by Aum/Yakuza/KGB.
- Will occur in the first minutes (10) of the strategic strike, following psychoenergetically crew-disabling QP weapons strikes.

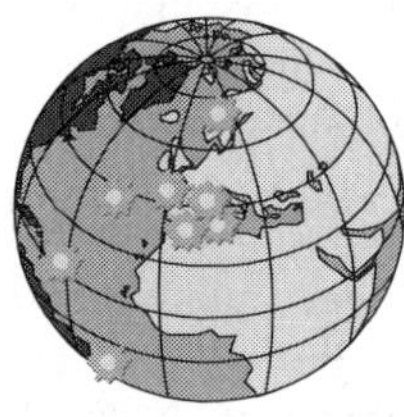
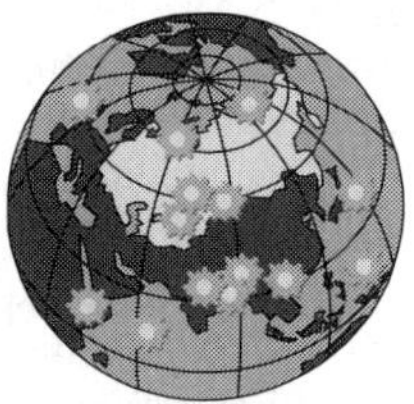

Note: Duds both overt and covert nuclear weapons and weapons-grade materials -- including covert weapons inserted on site in Russia which otherwise provide dead man fuzing.

Negative energy EMP

- Employs negative energy electrons (persistent Dirac Sea holes) and their negative energy EM fields and potentials.
- Extreme electromagnetic pulse of these negative EM energy fields strikes target with giant pulse of Dirac Sea holes, "eats" free electrons.*
  - All electron currents in target extinguished instantly.
  - Personnel die instantly; no nerve or brain currents. No cell twitches again; all bacteria, viruses, etc. killed. Soviets tested on two Afghan villages in 1980s war.
  - Mind-body coupling mechanism instantly destroyed, hence nickname "Mindsnapper". Mujahedin called it "wind of death".
  - All communications, electronics, missiles, nuclear warhead electronics, guidance systems, aircraft, ships instantly killed.

* Produces lattice positrons (excess positive charges) by eating orbital electrons.

Negative energy EMP (2)

- Kill of an entire naval attack force, massed armor attack force, satellites, ICBMs, aircraft, tactical missiles, artillery shell fuzes, communications, tanks, ships, etc.
- Can kill an entire large installation or city the same way, instantly.
- Or can kill more localized targets such as electrical power plants, power distribution lines and substations, refineries, chemical plants, port installations, Pentagon, C2 facility.
- Can "eat" electron charges in a strong storm front, reducing the storm energy.
- All purpose weapon; second only to QP weapons, new generation of carrier robots with energetics weapons, and sleeper mass conversions.

Briefing to N-4 in 1996

- J-6 Staff Director for C4 systems (Adm. Cebrowski) requested briefing on Tesla electrodynamics. Naval command in Florida proposed this author brief the Admiral. I agreed.
- Naval Ordnance Lab and Naval Research Lab then set up a 40-man Ph.D. board as a gauntlet, directed I be interrogated and savaged by board.
- I refused. These folks had fiercely savaged UWB radar pioneers such as Harmuth and Barrett and claimed UWB radar was impossible.* They knew nothing of higher group symmetry electrodynamics; and were part of the problem, not the solution.

* They later claimed to have been "pioneers" of UBW radar themselves!

Briefing to N-4 in 1996

- N-4 informed "briefer has declined the briefing". Questioned why, so command had to tell him.
- N-4 immediately dissolved the gauntlet board, limited NOL and NRL to one member each at the brief, ordered that briefer not be interrupted.
- I then briefed the N-4 on what Tesla EM was, and strongly urged another weapons briefing on Russian energetics weapons. Second brief was not held.
- Later, three colleagues and I played intense, desperate role in Israelis countering and blocking a massive KGB strike on the U.S. by those energetics (Tesla) weapons, that would have destroyed this nation on May 1, 1997.

AUM SHINRIKYO: IF ANYONE HAD LISTENED AND UNDERSTOOD*

- Cult formed in 1987. Over 60,000 members. More than $1 billion in gold and cash.
- Taught that U.S. would strike Japan with devastating blow in latter 1997
  - Japan should make preemptive strike first
  - AUM would get mass destruction weapons
  - Was developing BW weapons (anthrax, botulism, "Q" fever, etc.)
  - Developed own chemical weapons
- Made sarin attack on Japanese subways
- Formed alliance with Japanese Yakuza

AUM SHINRIKYO
"SUPREME TRUTH" CULT

Shoko Asahara, Leader

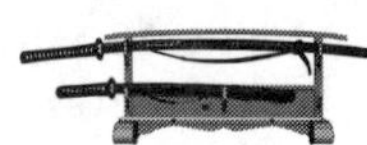

AUM SHINRIKYO: Alliance with the Japanese Yakuza

AUM SHINRIKYO
"SUPREME TRUTH" CULT

Shoko Asahara, Leader

- Formed alliance with Yakuza
- AUM/Yakuza leased operational use of strategic Soviet energetics weapons (strategic scalar interferometers) in latter 1989
- Trained crews, planned to strike U.S. a terrible "preemptive" blow in 1997 by assisting KGB
- Using deception, killed targets such as TWA-800
- Shoko Asahara's public statements foreshadowed Yakuza/Aum Shinrikyo participation in the KGB's strategic plan to attack and destroy the U.S. in 1997.

Asahara's public statement

- Asahara's public statements hinted at his plans with the Yakuza to participate in the scheduled FSB/KGB strike on the U.S. in 1997.
- Quoting Asahara:

  "I am certain that in 1997, Armageddon will break out. By 'break out' I mean that war will erupt and that it will not end soon. Violent battles will continue for a couple of years. During that time, the world population will shrink markedly... A Third World War will break out. I stake my religious future on this prediction. I am sure it will occur."

Kill of TWA Flight 800 on 17 July 1996 by Yakuza/Aum Shinrikyo crews

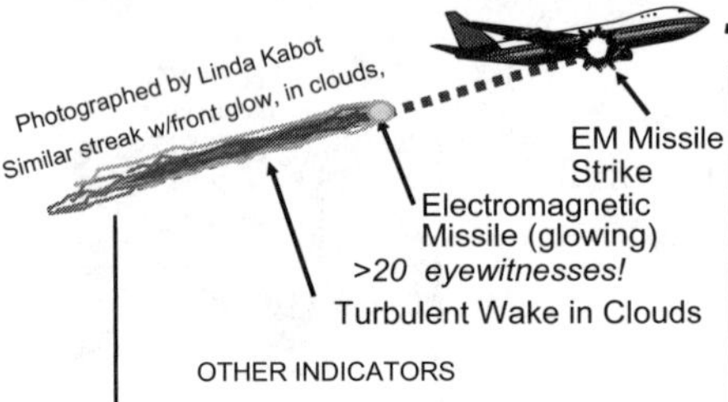

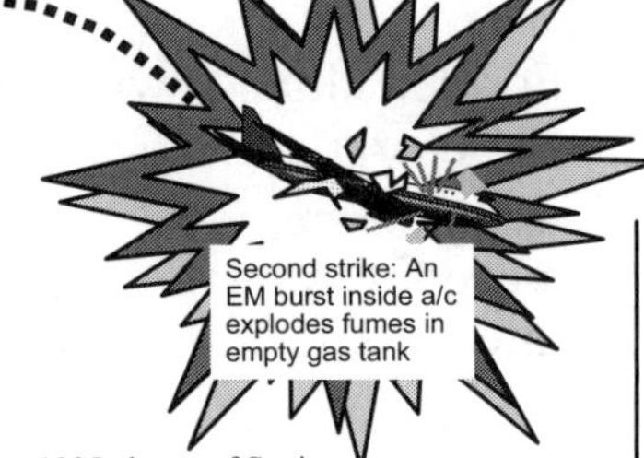

OTHER INDICATORS

- Helicopter eyewitness to streak, initial strike glow, and then larger second glow.
- Pilot eyewitnesses, multiple aircraft.
- Radar tracked a nonsolid blip that closed with the aircraft.
- Explosives ruled out.
- Significantly differs from downing by a light SAM at envelope of its range.
- Simple mechanical failure does not fit.

Emulation of surface-to-air missile (part of deception plan)

- 1985 photos of Soviet EM missiles offset from shuttle launch.
- 1985 eyewitness of similar EM missile strike on Arrow DC-8 at Gander AFB, Newfoundland.
- Multiple photos of EM missile tests over Western Australia (range).
- Probable: U-2 explosion 7 Aug 96.
- C-130 crash 22 Nov 96.

Emulation of a "leaking fuel" explosion (deception plan)

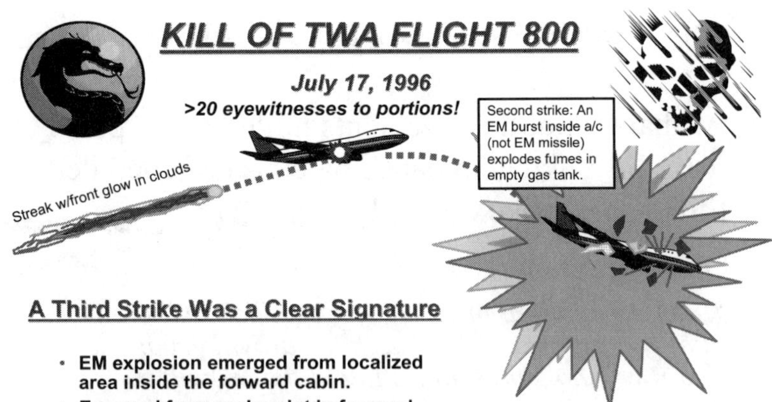

A Third Strike Was a Clear Signature

- **EM explosion emerged from localized area inside the forward cabin.**
- **Emerged from each point in focused spacetime inside some of the bodies.**
- **Bodies exploded from inside out.**
- **Bones blown right out through flesh, etc.**
- **Very different from explosion damage to a body from outside in.**

EM MISSILE PRACTICE

Latter Sept. - Early Nov. 1996

[Many Witnesses]

Melbourne, Australia
Australian Associated Press
Nov. 2, 1996

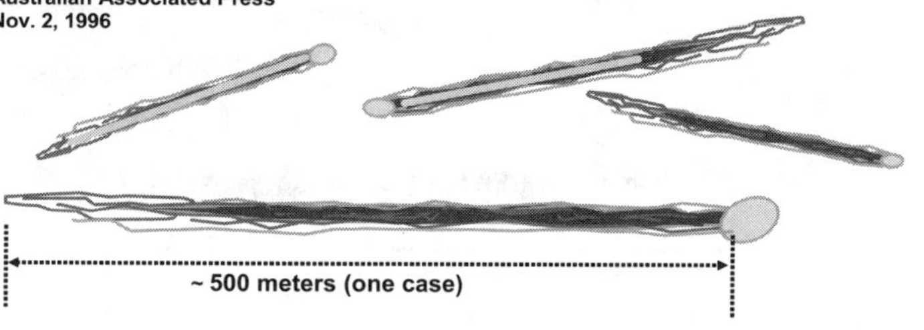

Aum Shinrikyo/Yakuza also conducted extensive firing practice tests over Western Australia. Tests after TWA-800 demonstrated what had killed the aircraft.

EM MISSILE PRACTICE AND DECEPTION

Nov. 17, 1996 at 2220 hrs.

Speed Bird 226

Lufthansa 405

Vicinity of Long Island, New York
Not far from where TWA-800 was killed
Two aircraft sighted object

Note: 4-month anniversary of kill of TWA-800. "Inside gloating" by Aum/Yakuza/KGB.

Probe to see if U.S. Intelligence recognized what had happened to the TWA-800. They didn't.

EM Missile Practice
Latter Sept.-Early Nov. 1996

(Multiple Nights, Multiple Streaks Each)

(Many Witnesses)

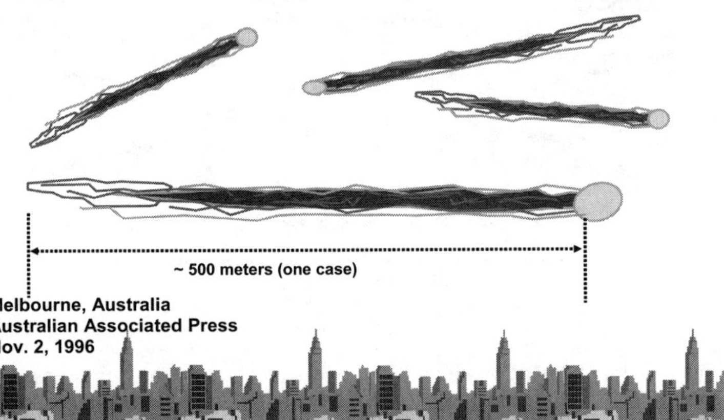

(SLIDE BRIEFING CONTINUED)

PART D: Second and Third Attacks Scheduled and Aborted

Second and third attacks scheduled and aborted

1997 was a very busy year

Section contents

Attack countering in 1996-97

- Contacted by Harry Mason, Australian geophysicist.
  - First-hand information on anomalous lights, explosions, quakes in West Australia.
  - Obviously were energetics weapons tests.
  - Alerted me to involvement of Aum Shinrikyo, Yakuza.
  - Furnished material on Aum Shinrikyo and Yakuza.
- Mason eventually visited me here in the U.S. for illuminating discussions.
- From these, we pieced together modifications to Soviet 1996-97 attack scenario.
- Had it not been for Harry's input, we would not have had the picture in time to alert the countering nation.

Actions 1996-1997 (cont'd)

- Mason provided special information that consortium of Aum Shinrikyo and Yakuza had leased use of the earlier generation Soviet energetics weapons.
- Initial downpayment $900 million in gold.
- Obtained Sen. Sam Nunn's report on terrorism, etc. including Aum Shinrikyo. Researched Yakuza.
- Feverishly increased work on weapons briefing.
- By late 1996, three colleagues and I desperately were trying to get message to high Israelis, that attack was forthcoming. Israel was and is the only nation which could and can counter such an attack by a rational foe.
- FSB/KGB attack first scheduled for February 1997.

Desperate actions

- Attempted to alert former Israeli ambassador to UK, who had confirmed weapons to UK intermediary.
- Ambassador was assassinated one week before emissary could reach him.
- In desperation, sent alerting fax to Israeli Embassy Defense Attaché, in Washington, D.C.
- A well-known Ph.D. and close colleague wrote a supporting study. Another close Ph.D. colleague delivered a copy to the Israeli Consulate in California.
- Scheduled attack countered by Israeli action; aborted.
- Informed U.S. channels of actions (not believed).
- Relaxed, breathed easier, back to energy and medical research.

Attack rescheduled

- To our astonishment, shortly it became obvious that *the attack was rescheduled for May 1, 1997*.
- We were completely perplexed as to how that could be, and Russia *not* disappear from the Earth if the attack were launched.
- A great deal of anguish and wracking of one's brain resulted.
- The answer was disarmingly simple, so it was difficult to recognize.

Answer: Israelis were sane; but the problem was insane

- Israelis had incorporated a *sane* C2* system.
  - Wait till recognize preparations, meet, decide, then issue launch order for preemption.
  - Easily countered by negative energy EMP.
  - Kill all QP personnel suddenly, then destroy site.
  - Counter simply by rescheduling order of attacks.
- Must be *insane* C2 system.
  - Every 2 hours, computer issues real launch order.
  - Launch sequence starts "for real", threat or no threat.
  - A time interval is available in the sequence where code can be entered to abort attack and revert weapons, if crews are alive.
  - Otherwise, QP weapon systems fire and Russia disappears from the face of the Earth. In short, it is a wrinkle on the old dead-man fuzing and mutual assured destruction (MAD).

*Command and control.

Desperate situation, desperate action

- Latter April; clock ticking, five days from attack and certain destruction of the U.S.
- Desperately faxed Israeli Prime Minister directly, explaining problem and solution.
- Urgently requested Israel switch C2 to insane system immediately, and notify Russians.
- It was quickly done.
- Two days before attack, the armada began standing down, much to the relief of my colleagues and me.
- Informed official U.S. channels of all actions. They did not believe it.

A bit of irony

- At about the time the great armada started standing down, Secretary of Defense Cohen was making his 1997 statement at a Georgia conference, confirming the earlier scalar interferometry weapons.
- The Secretary did not know that the attack had been imminent, or that it was standing down as he spoke.
- The U.S. Intelligence Community still did not believe it at that time, and did not know of the attack plan.
- By a hair's breadth the United States survived, thanks to the friendly nation and its QP weapons.
- The FSB/KGB simply resumed its planning and maneuvering toward another attack opportunity.

Afterword

- Also warned Israel and others about mind control capabilities achieved by Russia, evidenced by test kills of Captains Button and Svoboda.
- Some time later (unsure when), the main facilities producing those weapons were destroyed.
- Those facilities will remain destroyed for a great many years; but others now operating elsewhere.
- This pre-emptive action gave us a couple years of respite, until the robot attack at the end of 1999 occurred as a special "Millennium Gift" to the U.S.

Afterword (2)

- FSB/KGB strategic plan to attack and destroy United States must include nullifying Israeli counterstrike by QP and other major superweapons.
- FSB/KGB plan of Strategic Asymmetric War must include such neutralization.
- Major neutralization possible by use of massive numbers of "sleeper" conversion personas inserted into (a) appropriate governmental leaders and into (b) control/operational personnel for key superweapons.
- Sudden massive conversion in Israel disrupts government, produces utter chaos. Israel destroys itself.
- Asymmetric war passed into Operations Phase circa beginning 2005; means massive sleeper conversions being inserted worldwide, ready for triggering.

World's first subspace *precursor robot* war

An eerie war fought *inside* internal structures of EM fields and waves. I.e., fought inside group internal space (a special subspace).

Section contents

- **KGB continues developments.**
- **Aum Shinrikyo and Yakuza involvement.**
- **Weather engineering.**

- **Causal system robots.**
- **Precursor robot tests.**
- **Subspace precursor robot war.**
- **Afterword.**

FSB/KGB continued weapon development and attack planning

- Incidents and tests continued worldwide.
    - By FSB/KGB.
    - By rogue Japanese (Aum Shinrikyo and Yakuza).
    - By other nations.
- Several nations/groups (including Yakuza) used Western Australia for a target range.
- Extensive tests documented by geophysicist Harry Mason.
- Occasional aircraft, missile interference or kill.
- Full-bore weather engineering continued.

Aum Shinrikyo and Yakuza involvement

- On site in Russia, manning strategic scalar interferometer sites (kind that killed U.S.S. Thresher).
- Heavily engaged in weather warfare over the U.S. and North America since early 1990, other range practice in Australia.
- Itching to begin firing to destroy U.S. cities and major installations.
- Yakuza now a major, hidden world force.
    - Penetrates all major Japanese companies.
    - Penetrates Japanese government.
    - Large presence in oriental populations in U.S. cities.
- Yakuza is FSB/KGB protégé for asymmetric warfare against the internal U.S.; intends to destroy us.

U.S. weather engineered by rogue crews on-site in Russia (1)

- Steering jet streams.
    - Make low pressure areas (endothermic).
    - Make high pressure areas (exothermic).
    - Move highs and lows in coordinated fashion, to entrain and steer the jet streams.
- Displace cold air from Canada, warm air from Gulf.
- Where they meet provides sharp fronts and violent weather.
- Basic "cross-ploughed field" grid pattern covers a broad area for structuring in the grids (choosing localized cooling, heating, force engines in air.

U.S. weather engineered by rogue crews on-site in Russia (2)

- Placing sharp bends in jetstream adds great angular momentum.
  - Spawns tornadoes, powerful storms.
  - Generates much atmospheric energy and lightning.
  - Ignites fires in those areas deliberately kept dry.
- Maneuvering and pulsing high and low pressure areas vertically above each other can produce sudden downbursts.
- Adding spiral twists can generate sudden tornadoes.
- Created earthquakes and stimulated volcanoes into eruption; generated tsunamis.

Causal (precursor) robots and the subspace war of 1999-2000

- For any set of force dynamics in a material system, there is a corresponding set of nonmaterial precursor dynamics.
- That set is a nonmaterial precursor "engine" (cause) for the physical robotic system's dynamics and functioning.
- The precursor engine can travel through and inside Whittaker structure of an EM signal, wave, field, potential, etc. Easily travels through matter and through EM radiations from target.
- A material robot thus has a precursor robotic "engine".
- When this precursor robot engine interacts with proper type of matter, the material robot functions emerge and are performed.
- FSB/KGB developed formidable army of the robots.
- Unleashed them throughout the U.S. in latter 1999. *Israelis destroyed precursor robot army in latter 1999-early 2000*.

Causal System (Nonmaterial) Precursor Robots

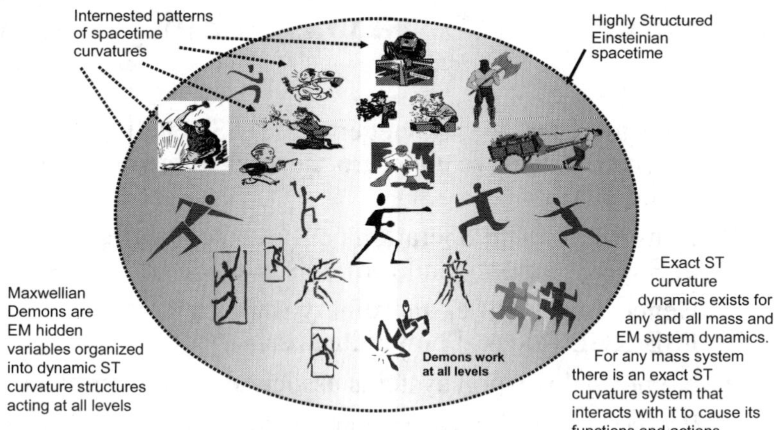

Causal system robots (CSRs). These are sets of ST curvature engines, with functions organized into a single complex weapon system with communication, propagation, weapons effects, and command and control functions incorporated. When the CSR interacts with targeted mass, the mass and its fields perform those system functions.

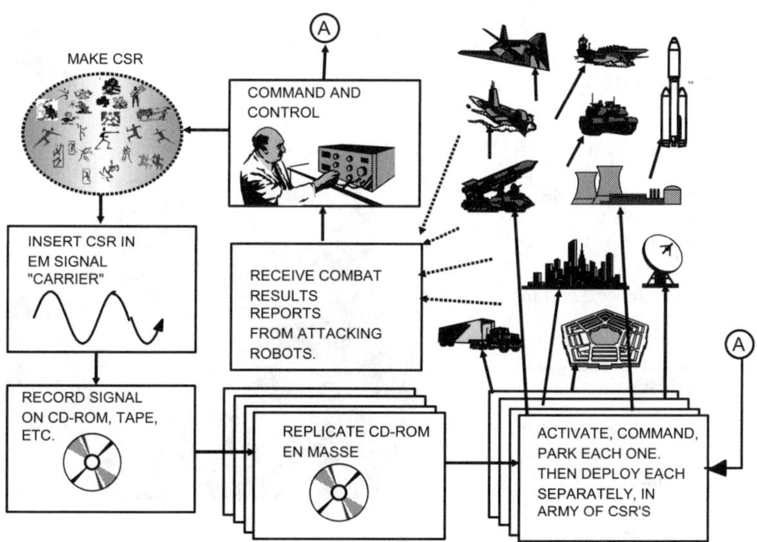

Mass reproduction of a CSR, once made, is simple and inexpensive.
The CSR resides in, or propagates inside, any EM signal, potential, or field.

Possible CSR* weapons complexes

- Prior to collapse of Soviet empire, U.S. satellites observed extensive underground sites in preparation (hundreds).
- Construction and operation continued even after collapse of Soviet economy and former Soviet Union.
- Hence must be of extraordinary importance, and to FSB/KGB (the real power in Russia).
- No normal weapon systems associated with complexes.
- Probable sites for research and development, deployment and worldwide employment of CSR robots.

---

* Causal System Robot. Same as nonmaterial robot; vacuum engine robot, spacetime curvature robot, precursor robot. See slide in Appx 1 on cause and effect.

Prior to planned robot "millennium" attack at end of 1999, first few days of 2000

- Stimulating apparently the only analyst capable of recognizing a CSR attack and connecting it with the scheduled giant strategic CSR attack.
- KGB General commanding the energetics weapon program previously confirmed that what this analyst writes on his computer is received as soon as he writes it (probably by Tempest techniques or energetics).

- Present author was struck by CSR in latter 1999 to give a strong but deliberately nonfatal stimulus. Reported; intelligence agencies did not believe it.
- If planned attack had occurred, Western analysts would have believed it was a severe Y2K problem, more serious than had been anticipated.

Robot attack on this analyst as an intelligence probe in latter 1999

- Computer suddenly controlled itself. Screen split, with two horizontal bands about two inches wide each, one at top and one at bottom.
- Light thus entering eyes from two radiant sources (scalar interferometry). Engine appears in interference zone.
- Sudden violent rapid flashes, as if 100 flashbulbs at once. <u>*Created in vision center where nerve signals crossed*</u>.
- Instant heart fibrillation, nervous system malfunctioned. Collapsed onto keyboard.

LAST QUARTER 1999

- Ten seconds exposure (30 would be fatal).
- Required 4 days to recover. Then we notified the Israeli Military Attaché in Israeli Embassy, etc.
- Israelis destroyed millions of robots throughout America.

Massive nonmaterial robot attack 1999-2000: First subspace war

- Y2K problem ongoing, tremendous work ongoing to avoid it.
- Millions of CSR robots were sent in by the KGB* through U.S. EM signals, and through EM fields and potentials in matter of ocean and Earth.
- Penetrated all over U.S., in satellites, command centers, ICBM electronics, nuclear sub electronics, computers, NSA, CIA, DIA, etc.
- Struck ("pinged") this analyst, to watch him notify "system" and see if the official system understood.
- We notified U.S., but we also notified Israel.

*New name is Federal Security Services, but the leopard has not changed its spots. Superweapons controlled by old die-hard Communist faction, who wish to destroy the U.S. Younger faction (from which Putin came) wishes accommodation. Unfortunately the old faction is in control.

 Earth's first subspace war 

Ace in the hole

- Israel quickly destroyed millions of robot weapon systems on site in electronics and installations and computers and command centers in the U.S., in Israel, and in other Western nations.
- A strategic war of eerie kind occurred in those few days, and it did not make a whimper or attract attention of our intelligence community. Millions of robots were destroyed.
- The "infolded" EM inside all normal EM fields, potentials, and waves is a special "subspace" mathematically, including inside the Earth's matter and the ocean's matter.
- <u>The first great subspace war was fought at the change of the millennium for several days, inside EM signals in our systems, computers, etc., and no one knew it.</u>

Afterword to the eerie 1999-2000 strategic war

- The West slept on and did not believe the eerie nonmaterial precursor robots – or the war – at all.
- We also warned the "little nation" of the mind-control psychoenergetics tests as against Button and Svoboda, and the coming capability of this threat.
- The subject had arisen in the previous attack scheduled for May 1, 1997, which was narrowly avoided.
- Many months later a garbled communication indicated that Israel had suddenly struck those psychoenergetics facilities also, totally destroying them.
- We did receive an "attaboy" for in-the-nick-of time warning provided in latter 1999.

Countdown toward destruction

Converging WMD threats

Section contents

- **International terrorism.**
- **Wake-up call.**
- **Subsequent events.**
- **Calm before the storm.**
- **Personal conclusions.**
- **Time criticality.**

International terrorism escalates

- International terrorists struck hard on Sept. 11, 2001.
  - World Trade Center (Twin Towers) and Pentagon.
  - One aircraft, redirected, was targeted against White House and the President. Crashed by onboard struggle.
  - Osama Bin Laden was a household word.

Twin Towers burning after strikes by two aircraft - AP Photo.

- Suddenly the "D.C. Doldrums" regarding terrorism and WMD threats were destroyed.

The wake-up call

- A month or so later, President Bush and VP Cheney had been rather thoroughly briefed.
- They understood the nature of this war. VP Cheney said:

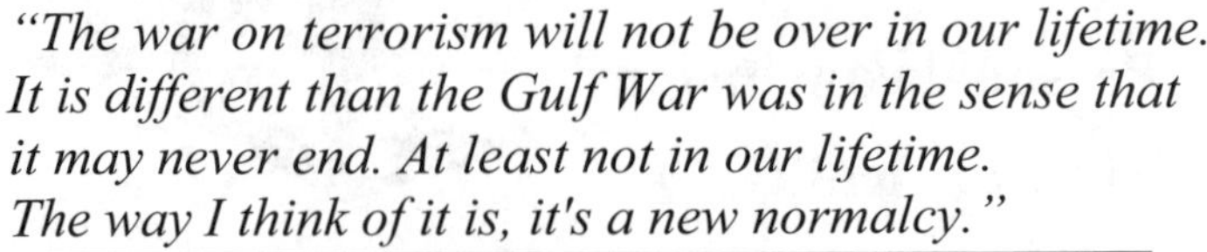

"The war on terrorism will not be over in our lifetime. It is different than the Gulf War was in the sense that it may never end. At least not in our lifetime. The way I think of it is, it's a new normalcy."

Vice President Dick Cheney, October 21, 2001.

???

Subsequent events

- President Bush retaliated against terrorism.
  - Afghanistan, where the Taliban had long harbored Osama Bin Laden.
  - Iraq, with its record of backing terrorism and mass murder of its own civilians (and its capability for WMD).
- Occupation of Iraq and Afghanistan is ongoing, as U.S. strives to get free, modern governments installed, economy restored, and proper internal security forces formed and trained. Resistance and incidents continue. Bin Laden remains at large, Saddam Hussein is caught and awaiting trial. Hussein's two sons have been killed.
- For FSB/KGB, this is a "bleeding the U.S. dragon" period.

Calm before the storm

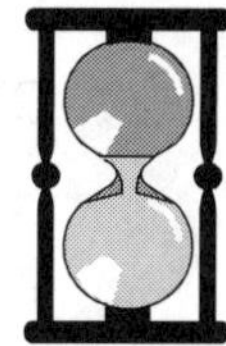

- U.S. national election also sowed divisiveness and preoccupied nation for more than a year.
- International terrorists continue insertions.
- Yakuza prepares portable interferometers; North Korea and Iran pursue nukes.
- Castro's inserted 10,000 guerrillas still here.
- Russian nuclear weapons still in U.S. cities.
- Chinese quiet insertion probably ongoing.
- Russian KGB has prepared robots again.
- *Inserted conversions finishing*; probes will be evident as buildup almost certainly underway inside the U.S. itself.

Personal conclusion of this analyst

- Multiple strategic WMD threats and attacks are rapidly converging on the U.S. simultaneously.
- U.S. woefully unaware of the eerie war and capabilities.
  - U.S. in grave danger, may be destroyed.
  - *U.S. scientific community has failed us*.
  - Government and intelligence community have totally inadequate scientific advice.
  - Meanwhile, clock ticks toward midnight.
- Insertion phase completed; now in operations phase.
- Yakuza registered superweapons on Yellowstone Caldera; achieving knockout capability against U.S. with one shot.
- Shot 9.3 seabottom quake on Dec. 26, 2004 which caused giant tsunamis in Asia.

---

* T.E. Bearden, "Errors and Omissions in the CEM/EE Model", at http://www.cheniere.org/techpapers/

Threats - working

- **Energetics weapons threat.**
- **Bioenergetics weapons threat.**
- **Robots weapons threat.**
- **Psychoenergetics weapons threat.**
- **Conversions threat (to be discussed).**
- **Terrorists nuclear and BW weapons threat.**
- **Chinese threat.**
- **Yakuza/Aum Shinrikyo threat.**

(SLIDE BRIEFING CONTINUED)

PART E: Bioenergetics and Its Weaponization

Bioenergetics and its weaponization

Use of infolded EM against living bodies, biofields, and biopotentials

Section contents

- **Basic bioenergetics.**
- **Selected tests.**
- **Body radiates its internal patterns.**
- **Information content of the field.**
- **Ramifications.**
- **Kaznacheyev experiments and results.**
- **Microwave radiation of U.S. Embassy.**
- **Becker's work and ramifications.**
- **Prioré's work, results, and ramifications.**
- **Shadow disease state, immune system reaction.**
- **Spreading the immune system.**
- **Gulf War Disease: Russian test.**
- **1998 proposal for portable treatment unit.**

Basic bioenergetics

- Scalar interferometry.
- Longitudinal EM waves.
- Bioenergetic Engines.
- Negative energy EMP.
- Quantum potential.
- Nonmaterial robots attacking bodies.
- Carrier robots with bioenergetic weapons.

Selected bioenergetics tests

- Kaznacheyev experiments
- Microwave radiation of U.S. Embassy
  - Disease induction
  - Deaths of three U.S. Ambassadors
- Gulf War Syndrome
- Immune system spreading
- Carrier robot test against this author

The body continually receives and retransmits EM energy

- The body may be considered as a dielectric.
- Every part of the dielectric participates in absorption and emission of each photon to and from any point on the surface.*
- Mechanical force effects are also electrical in nature and exchange similarly.
- CEM/EE does not account for precursor engines in the exchange.

* Reali, G. C., "Reflection from dielectric materials," American Journal of Physics, 50(12), Dec. 1982, p. 1133-1136. The reflected field from a dielectric material is not generated just at its surface but comes from everywhere in the interior of it.

Ramifications (1)

- An EM "dielectric transmission path" connects every part of every cell to every part of the body surface.
- Electrical processes in the body, including deep within the cell and its nucleus, participate in absorption and emission of every photon to and from the body's outer surface (and every mechanical force experience).
- All body electrical processes participate in exchange of precursor engines between body and environment.
- EM emission energy of the body is a conglomerate mix of fractions of all processes and functions ongoing in the body or affecting it, including in every cell. Including the engines.

Ramifications (2)

- Reaction of each photon absorbed on any point of the outer surface of the body, connects through the dielectric transmission path to every part of the body, including to every part of every cell*. Skin absorption connects incident signals and engines everywhere throughout body also.
- Every body emission is participated in by every part of the body, including all cells.
- Body emission "changes" environment, adding EM radiation and engines.
- Body's EM absorption allows environment to change or affect every internal function and part of the body.
- The body's reaction (healing) system continually readjusts, corrects deviations from normalcy, in every cell.

* In quantum physics, part of the photon is localized and part is nonlocalized, reaching even across the universe.

Ramifications (3)

- Total of all impinging external EM radiation upon the body, plus mechanical "special EM" radiation, affects and changes body and all its processes, participating in them.
- Impinging EM radiation engines can be manipulated to either harm or heal the body and any or all of its cells.
- The body must continually overcome these externally-induced EM changes in its cells and processes, to maintain normalcy.
- The poorly studied *cellular regenerative system* performs the restorative function, which is the fundamental "healing" function of the body.*

*The cellular regenerative system was mostly studied by Becker and a few others. See R. O. Becker, and David G. Murray, "The electrical control system regulating fracture healing in amphibians," Clinical Orthopaedics and Related Research, No. 73, Nov.-Dec. 1970, p. 169-198.

Ramifications (4)*

- EM fields and waves decompose into differential structures of two scalar potentials.
- Each of the scalar potentials decomposes into a harmonic series of bidirectional longitudinal EM wavepairs (LWs). "Engines" (special LW sets) may be inserted and carried.
- Zero vector summations of such internally structured fields make scalar stress potentials with infolded structured longitudinal wave structures and dynamics.
- Correct multiwave EM waves radiated onto body pass or diffuse back through all "emission channels" (portholes).
- Any internal part or function of cells, etc. can be directly affected, changed, controlled, etc.**

---

* See the two Whittaker papers, 1903 and 1904, referenced elsewhere.

** See also Porthole Briefing on website www.cheniere.org, as well as Provisional Patent Application freely given to public domain.

Devyatkov's "information content of the field" (engines by another name)

- Nonthermal millimeter waves.
  - Used in living systems.
  - Sharply resonant actions.
- Hypothesis now a theory.
  - Experiments since 1960s.
  - Confirmed.
- Now basis for
  - Experimental research.
  - Use in medicine and biology.
- New kind of photobiology.

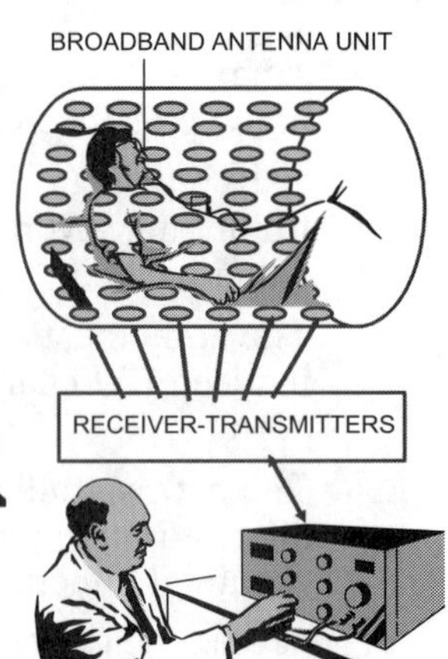

Information content of the field (2)

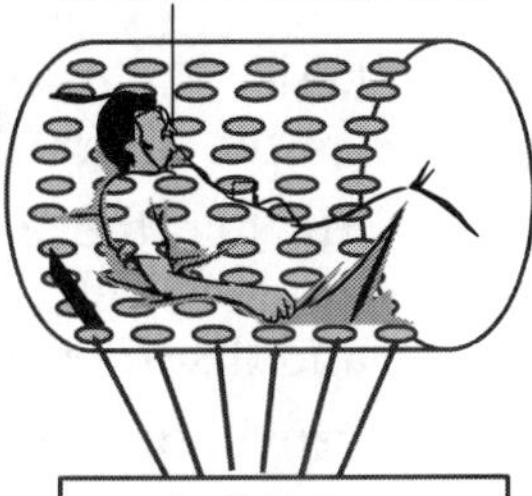

- Contrary to communications theory
  - Effect increases as frequency *decreases*
  - *Number* of resonant frequencies determines
  - Signature of time-component engines
- For a resonant system
  - Average linear dimension L, wavelength λ
  - V = propagation velocity
  - $\lambda = V/f$
- Number of resonant frequencies
  - Increases as λ decreases
  - In proportion to L/λ
- For each resonant frequency:
  - Particular distribution in the system
  - Determines state of the system

Extended dielectric radiation properties

- Every part of body dielectric participates in absorption and emission of each photon at any point on the body's surface.* The energy portion of the photon carries engines.
- The forces are due to ongoing interactions between precursor energy engines and the body's charged matter, at all levels from smallest to largest.
- All the body's internal EM fields, potentials, and waves decompose via Whittaker 1903 and 1904 into harmonic series of longitudinal EM biwave pairs and their dynamics.
- Thus all external EM radiation of the body carries the internal biwave pair dynamics of the total interior, including every cell and every condition.

* Reali, G. C., "Reflection from dielectric materials," Am. J. Phys. Physics, 50(12), Dec. 1982, p. 1133-1136. The reflected field from a dielectric material is not generated just at its surface but comes from everywhere in the interior of it. This results in direct communication of engines between all elements of the body and between the body and all elements of the environment.

Proof that the cellular condition is radiated from groups of cells

- A diseased, damaged, or infected cell culture will and does emit "engines" for its exact internal EM condition, hence the disease or damage "generatrix form" itself (in the internal structuring of the EM radiations).
- Further, these "state" emissions can couple into other targeted cells to produce disease and disorder in them.
- Proven in some 5,000 Russian experiments by Kaznacheyev.
- Demonstrated induction of cellular disease and disorder between EM-coupled but otherwise environmentally shielded cell cultures.
- Similar experiments have been successfully replicated in the West by Reid *et al.*, and by Popp *et al.*

(Glass or quartz)

Disassembled

Kaznacheyev's two-chamber apparatus

(EM Induction of effects of diseases, physical damage, nuclear radiation, toxins)

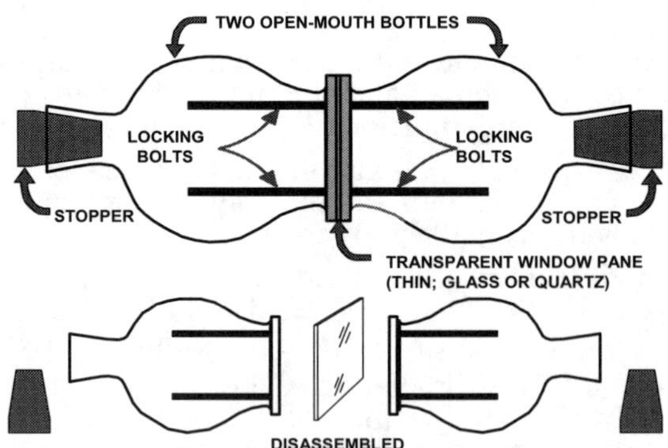

A cell culture is divided into two parts. Cells on the left are then damaged in some selected fashion. With a thin windowglass divider, cells on the right remain healthy. With a thin quartz divider, after a delay the cells on the right begin to die with the same disease, or exhibit the same damage. This directly demonstrates vacuum energy (subtle precursor energy engine) action.

Kaznacheyev's EM induction of cellular disease and disorder

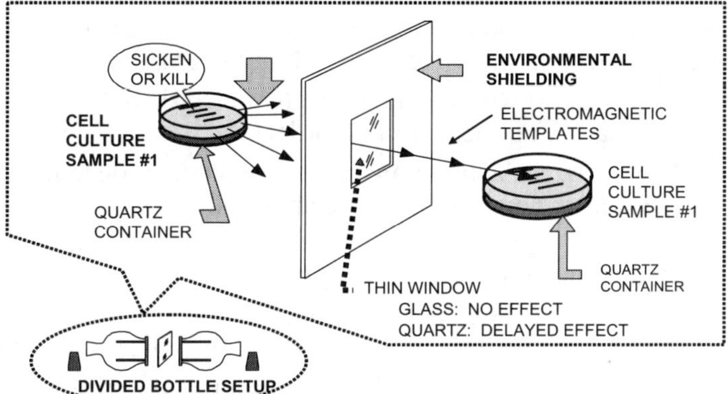

Note: Minimum lattice is one harmonic interval: IR to UV is such a minimum G-lattice. Consider particularly the dielectric pathway in the emitting diseased cells and in the receiving targeted healthy cells.

Another surprising feature of Kaznacheyev's experiments

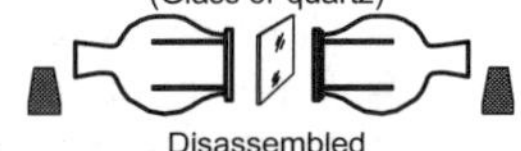

- Experiments do not work if targeted cells are in normal light, such as sunlight*.
- The normal light bandwidth occupies less than a harmonic interval between ultraviolet and infrared.
- Hence its difference frequencies also fill such an interval. For a set of n frequencies, the number *d* of difference frequencies is $d >> n$.
- The visible light spectrum "jams" specific internal EM signal inputs through dielectric pathway into the body, turning most environmental signals into harmless warmth.

* It has also been shown that if bacteria are killed in the "dark" by UV, then 24 hours (12 generations) later are placed in sunlight, a substantial fraction of the "dead" bacteria will revive. Suspended but maintained mind engines, matter engines, and their coupling engine after physical death are demonstrated. In other words, it demonstrates how living systems can be made and assembled from "parts". The Russian energetics weapons scientists have that full capability today.

Ramifications of Kaznacheyev's experimental results (1)

- At least one harmonic interval is necessary for the effect to be evidenced in the target cells. More than one is better.
- Interestingly, if we assume the body dielectric to be isotropically nonlinear (to first order):
- Velocities of the actual transmitted frequencies depend upon the particular point on the wave amplitude.
- Transmitted frequencies overshoot, interfere, breakup in nonlinear dielectric medium. *Difference* frequencies do not.
- Difference frequency passes through nonlinear dielectric medium like a sine wave passing through a linear medium*.
- West has not investigated direct communication of EM "hash" difference frequencies in dielectric human bodies.

* Owen Flynn, "Parametric arrays: A new concept for sonar," Electronic Warfare Magazine, June 1977, p. 107-112.

Ramifications of Kaznacheyev's experimental results (2)

- A heterodyne signal — at the difference frequency between an input signal and a reference signal — can be enhanced by adding noise* as is well-known.
- Thus the signal of the disorder in the cellular pattern emitted from diseased cells, received by normal cells, can be amplified electronically and rather easily.
- Powerline radiation (electronic smog, a form of noise) thus carries its own amplification mechanism. Eventually the amplification overcomes any squelching by sunlight, etc.
- The Russians promptly (as early as the late 1950s) weaponized and tested these and similar effects.

* Dykman, M. I. et al., "Noise-enhanced heterodyning in bistable systems," Physical Review E, 49(3), Mar. 1994, p. 1935-1942.

Microwave radiation of U.S. Embassy in Moscow

Former U.S. Embassy in Moscow

BACKGROUND

- Began in latter 1950s
- Discovered on VP Nixon's trip
- Initially thought to be *nuclear* radiation (Discovered with Geiger counter?)
- High level target — U.S. Ambassador
- Guarantees personal attention of:
  - U.S. Ambassador to USSR
  - U.S. President
  - NSA, CIA, DIA, NSC, etc.
  - Top consulting scientists
  - Leading U.S. scientific institutions
- Two U.S. Ambassadors died, another sickened
- Anomalous health changes in personnel, only in zero-field (zero potential gradient) areas!
- Four U.S. Presidents requested Soviets cease
  - Cut from 18 watts/sq cm to 2
  - Then again increased
- No one could understand what was going on
- Aluminum screens were placed over windows
- Moscow was declared a hazardous duty zone

More on Embassy radiation

- Diseases, health changes statistically significant, not accidental.
- Johns-Hopkins scientists measured *field* radiation patterns inside Embassy accurately.
- Health changes occurred only where *fields* absent.
- Concluded EM radiation was not the cause.* Their own measurements falsified their conclusion.
- Actually it proved some characteristic of the field-free (stationary) potentials was the cause, *due to their 100% correlation*. Direct indication of precursor engines.
- Standard electrical engineering erroneously considers the *force fields* in charged matter to be primary causative agents. They *are not*; the precursor force-free fields in space *are*.

* If EM radiation *were not* the cause, then the outbreaks would have appeared both in field areas and in areas devoid of fields.

Purpose of Embassy radiation

- Struck a high level target (U.S. Ambassador). Three U.S. Ambassadors died as a result. Hundreds sickened or had health changes.
- Ambassador target guaranteed attention of CIA, DIA, NSA, etc. Analysts baffled. Agencies turned to scientific advisors.
- *By U.S. reactions taken at the Embassy*, Soviets determined with 100% reliability whether U.S. knowledgeable in energetics and whether U.S. had defenses. They didn't. They *still* only know EE.
- U.S. put up window screens against the "normal EM field" radiations. Ineffective for stationary potentials.
- Proved our ignorance. Valuable intelligence assessment tool to the Soviets for decades.

RADU vs. Radio Free Europe*

- Mysterious radiation device, size of small watch
- Used by Romanian President to dispose of enemies
- Causes rapid growth of cancer in those persons exposed (per General Ion Papeca)
- Bizarre cancer deaths in Radio Free Europe
  - Four top officials
  - Included three successive Directors
  - After death threats from Romanian secret service
- Hence some other nations know of — and have — at least some of this technology.

---

*U.S. News & World Report, Dec. 12, 1988, p. 15.

Lunev* on Chinese capability (1)

- Lunev has cancer, as do most other GRU operatives that were stationed in China.
- Lunev is highest ranking GRU defector to U.S.
- When China and Soviet Union had basic disagreement, Chinese induced cancer in almost all the on-station GRU operatives.
- Lunev also states China is U.S.'s greatest threat.
- New alliance between Russia, China, and India.

---

* See Stanislav Lunev, Through the Eyes of the Enemy, Washington, DC: Regnery Publishing, 1998. Lunev was a Colonel in GRU intelligence.

Chinese capability (2)

- China has WMD and superweapons.
  - Nukes, ICBMs, subs, bombers.
  - Quantum potential weapons.
  - Negative energy EMP weapons.
  - Scalar interferometers.
  - BW weapons and bioenergetics weapons.
  - Probably well-along on psychoenergetics as well.
- China prepares both to reach accommodation with the U.S. and to be able to destroy the U.S.
- China's fiercely increasing energy needs are stoking the fires of U.S./China conflict, competing for foreign oil.
- China probably aware of the present FSB/KGB plan, use of the Yakuza, etc. Has new alliance with Russia.

Simplified diagram of the immune system

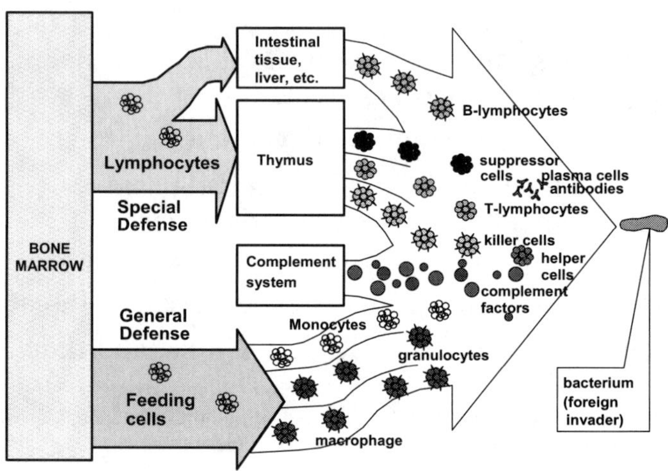

The puzzle of morphogenesis: Lack of knowledge of engines

- *"It remains a challenge, however, to understand precisely how the combination of tissue repair mechanisms with reactivation of embryonic programs can generate growth, pattern formation, and morphogenesis in an adult animal."*

Quoting Jeremy P. Brockes, "Amphibian limb regeneration: Rebuilding a complex structure," Science, 276(5309), Apr. 4, 1997.

Comment: Cannot be understood until a much better CEM/EE model is used for analysis and understanding, and until knowledge of how the master cellular control system and the cellular regenerative system use such electrodynamics with "engines" to cause forward actions (differentiation) and "anti-engines" to reverse damage (dedifferentiation).

Effects of dynamic structuring of the scalar potential: Engines

"It is now known that the d.c. potentials measurable on the intact surfaces of all living animals demonstrate a complex field pattern that is spatially related to the anatomical arrangement of the nervous system..."

Robert O. Becker, "The significance of bioelectric potentials," Bioelectrochemistry and Bioenergetics, Vol. 1, 1974, p. 188.

- ***Comment***: Potentials and their complex field patterns possess internal longitudinal EM biwave dynamics (engines) for all nervous system and cellular arrangement and functioning of the entire organism.
- The fields radiated from the body enfold samples of all such internal patterns and signals (all such engines).

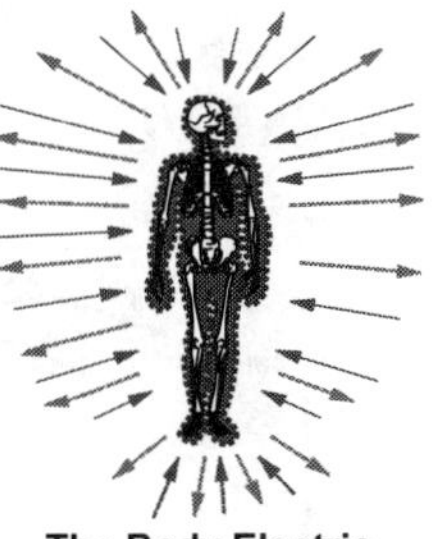

The Body Electric

Epochal work of Robert Becker

Controlled cellular dedifferentiation and redifferentiation, and the cellular regeneration system

Preliminary remarks

- Medical research and study centers on immune system.
  - <u>*Immune system heals nothing*</u>.
  - Fights invading pathogens, wins, cleans up.
- Healing and restoration accomplished by the <u>*poorly-studied*</u> cellular regeneration system.
  - EM system and EM techniques, not drugs.
  - Cannot be understood via standard CEM/EE. Requires higher group symmetry CEM to model it.
- Becker came within inches of deciphering the regenerative system; but CEM/EE failed him. Demonstrated cellular dedifferentiation and redifferentiation by EM means.
  - Nominated 3 times for the Nobel Prize.
  - Ruthlessly suppressed for his efforts. Funds pulled, forced to retire early.

Becker's proposed control system governing regeneration

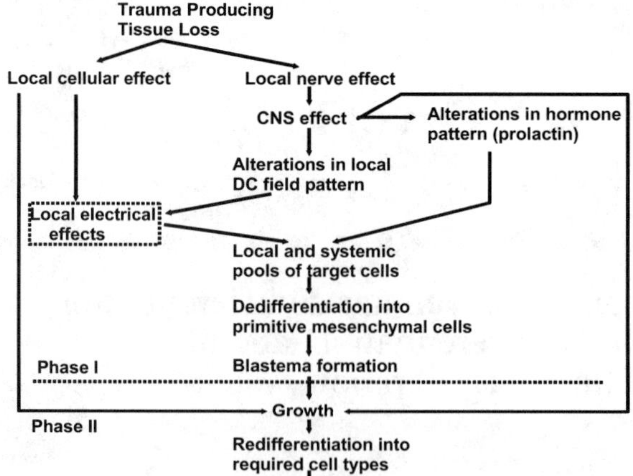

*Becker & Spadaro, "Electrical Stimulation of partial limb regeneration in mammals," <u>Bull. N. Y. Acad. Med</u>., 48(4), May 1972, p. 629.

Becker's theoretical control system involved with response to injury

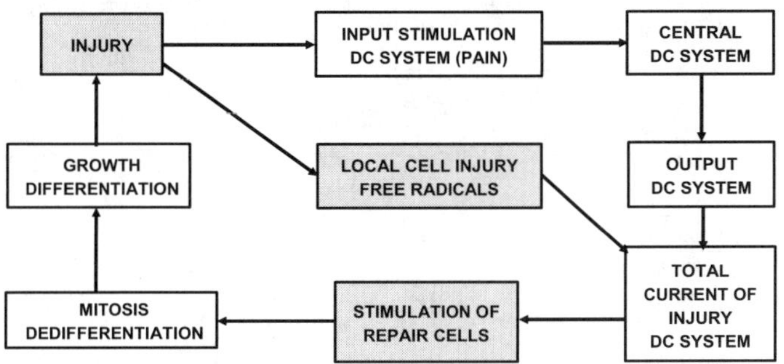

Note: Becker only had conventional EM available, hence was unable to model the inner structuring and inner dynamics of the fields and potentials. So he missed the deeper "engine and antiengine" mechanisms.

*Robert O. Becker, "The significance of bio-electrical potentials," Bioelectrochemistry and Bioenergetics, Vol. 1, 1974, p. 191.

Extending Becker's work

- Becker had only the archaic CEM/EE model available; flawed and grossly inadequate.
  - Insufficient group symmetry.
  - Already eliminates the "infolded" longitudinal wave EM that structurally and dynamically comprises normal EM potentials, fields, and waves.
  - Eliminates supersystem, engines, and anti-engines.
  - The missing internal LW EM is what is manipulated and used by the cellular regenerative system.
- Nonlinear phase conjugate optics not yet developed when Becker did his seminal work; time-reversed EM unknown.
- Successful higher symmetry electrodynamics models — such as O(3) — have been developed and are available.*

* See M. W. Evans, "O(3) Electrodynamics," Series Editor M. W. Evans, Modern Nonlinear Optics, Second Edition, 3 vols., Wiley, 2001; Vol. 2, p. 79-267

Author's diagram of cellular regenerative system functioning

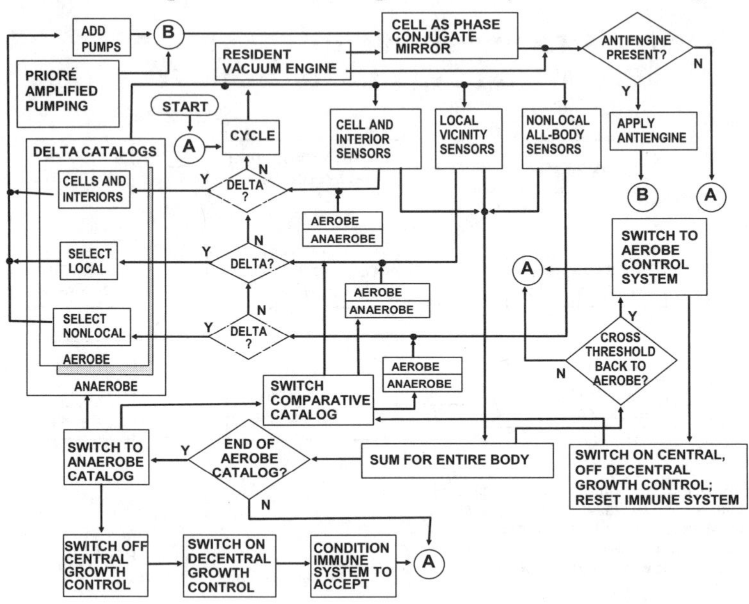

Becker's epochal bone fracture healing

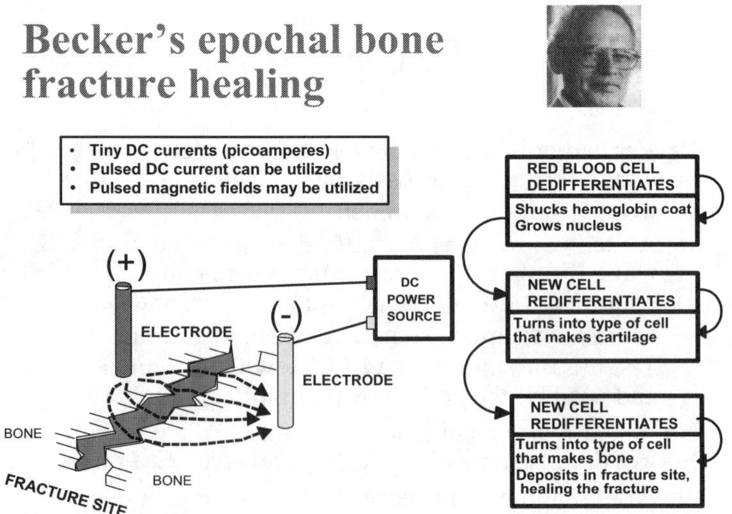

One of the few EM healing modalities approved by FDA and utilized in some U.S. hospitals every year to cure intractable bone fractures.

Prioré team's Bioenergetics: Revolutionary healing

● ● ● ● ● ● ● ● ● ● ● ● ● ● ● ● ● ●

Using infolded EM on living bodies, biofields, and biopotentials

Prioré's sensational results

- Revolutionary cures of terminal cancers and tumors in lab animals. Replicated in humans.
- Cured dread diseases such as trypanosomiasis.
- Reversed clogged arteries.
- Restored suppressed immune systems.
- Did double-blind tests, proving efficacy. Caused a sensation in French medical community.
- Some eminent French scientists worked on it.
- Spectacular results personally presented to l'Academie by Robert Courrier, head of biology section and Secretaire Perpetuel of the Academy.
- Independent tests by Courrier's personal assistant and by University of Bordeaux verified results

Prioré (right) makes a point to Chaban-Delmas, Mayor of Bordeaux

- Many of the Prioré team's experiments were done in Bordeaux.
- Antoine Prioré patents:
  - "Apparatus for producing radiations penetrating living cells," U.S. Patent No. 3,368,155, Feb. 6, 1968.
  - "Method of producing radiations for penetrating living cells," U.S. Patent No. 3,280,816, Oct. 25, 1966.
  - "Procede et dispositif de production de rayonnements utilisables notamment pour le traitement de cellules vivantes," [Procedure and Assemblage for Production of Radiation Especially Serviceable for the Treatment of Living Cells], Republique Francais Brevet d'Invention P.V. No. 899.414, No. 1,342,772, Oct. 7, 1963.

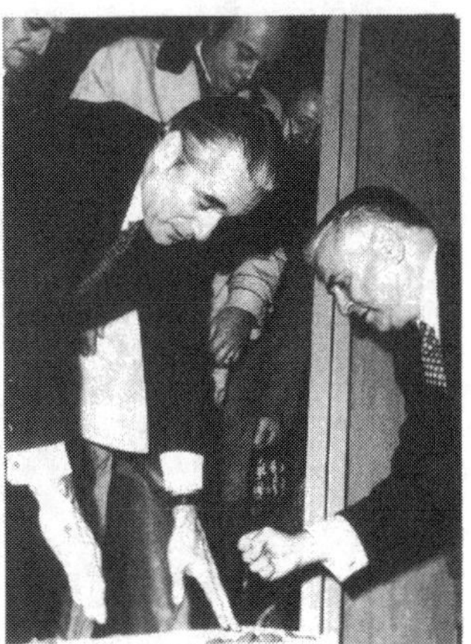

Prioré's laboratory and device for treating small animals

Mechanism of the Prioré therapy

- Produced specifically structured scalar potential
  - Some 17 EM frequencies were mixed in a rotating plasma
  - Phase conjugates added by plasma
  - Modulated onto a rippling magnetic field; carried it to every cell and all interior parts
  - Formed scalar potential with inner EM longitudinal pump waves
  - Hyperspatial pumping of cells/parts as pumped phase conjugate mirrors (PPCMs)
- Each disease has its "disease-specific" vacuum engine.
- Amplified anti-engine created by the LW pumping.

Mechanism of the Prioré therapy

- "Inner EM" pumping with longitudinal EM waves created amplified and <u>*specific anti-engine*</u>
- Amplified anti-engine rapidly reversed (dedifferentiated) cells/parts back to normal state
  - Cells returned to control by cellular control system
  - Restored cellular control system to full former functioning
- Scrubbed out the cumulated precancerous state
- Eliminated "shadow" disease state
- Restored the immune system to previous normal functioning
- No excessive trauma to treated animal

Block diagram of Prioré's healing (regenerative and time-reversing) process

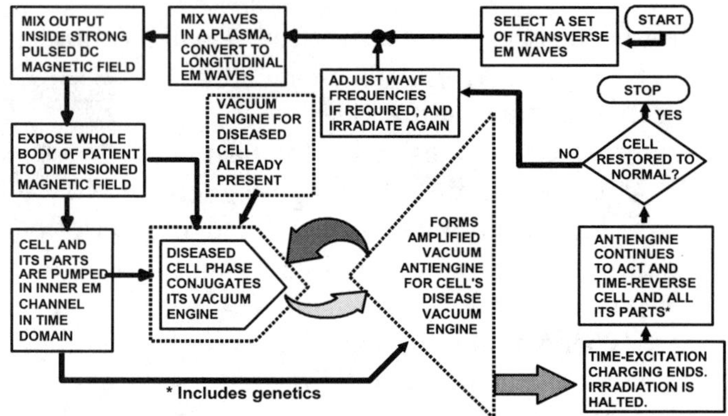

- The damaged or diseased cell has (1) normal engine and (2) delta disease engine.
- Making antiengine for both, and irradiating cell with it, time-reverses the cell and gradually erases the delta condition, healing the cell.
- This is the body's own cellular regenerative mechanism, which Prioré amplified.

Pautrizel's proof of time reversal of the stricken cells back to a previous state

- Pautrizel compared identical treatment of immature rats with immature immune systems, and mature rats with mature immune systems.
- Treatment restored the damaged *mature* system back to full *mature* functioning. It then recognized and dispatched the pathogens.
- Treatment restored the damaged *immature* system back to full *immature* functioning, unable to cope. The pathogens promptly reinfected and killed the immature rat.

PROF. RAYMOND PAUTRIZEL
(Renowned Parasitologist)

Pautrizel's proof of time reversal of the stricken cells back to a previous state (2)

PROF. RAYMOND PAUTRIZEL
(Renowned Parasitologist)

- **The lesson**: The body is indeed returned to a previous physical state, or *dedifferentiated* back along its previous differentiation pathogenic path through time.
- Pathogens themselves not killed by the treatment, but by restoring the ability of the immune system to recognize and destroy them.
- Only the *previous ability* of the immune system is restored by the process, which is a method for controlled biological dedifferentiation (time reversal of the stricken cells).

3-Stories high treatment facility

Prioré's large plasma tube treatment facility for humans. Results were duplicated in human experiments.

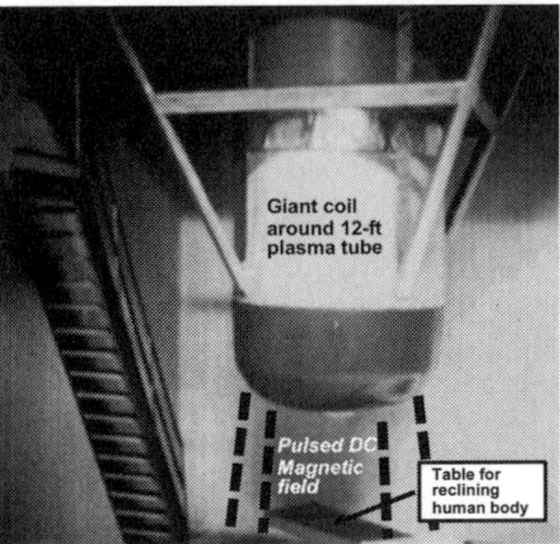

- At the end of the project, Prioré had developed a large unit to treat humans.
- A few were treated.
- Cures replicated what had been done in lab animals.
- The massive size of the unit and long treatment time (hours) made the process bulky and very inconvenient — and very expensive.
- We proposed a way to do it with a suitcase-sized unit and a treatment time of two minutes. (See *Porthole Briefing*, www.cheniere.org)

17-foot Prioré plasma tube

Used in large machine (3-stories high) to treat a limited number of human patients

Plasma tube converts transverse EM waves to longitudinal EM waves, impressing them inside a magnetic field in a huge coil around the tube. The "engine-carrying" magnetic field then penetrates the targeted living body down to atomic nuclei.

Some peculiar results by the Prioré team

- Once a rat is cured of a particular disease:
  - A single drop of blood from the cured rat, injected into another rat stricken with same disorder, would result in the injected rat getting well slowly.
  - Ability to transfer immunity and healing in the extracted drop of blood gradually decayed over period of a few weeks.
- Even the fields and potentials in the blood had been altered internally, in the treated animal.
- Injected animal's cellular regenerative system could time-domain pump over the entire body, everywhere reducing the delta between that injected drop of blood now the rat's own, and the rest of the blood in its body.
- Shows *anti-engines in molecular potentials* of the blood of the cured rat, and diffusion of this antiengine pattern in EM exchange with the environment as time passes.

Regeneration in a strain of rat with part of its immune system missing

- In mammals, the immune system plays a role in reducing the regeneration capability, compared to that of more primitive systems.
- In a strain of rat where that "suppressive" part of the immune system is missing, the regenerative capability is dramatically increased.
  - Can regrow cut away sections of tail.
  - Regrows plug cut out of ear.
  - Reconnects severed optic nerve.
  - Partially restores severed spinal cord.

- The suppressive "delta" engine of the immune system, operating simultaneously with the underlying primitive regenerative engine, sums to the modern much-reduced regenerative engine in the mammal.

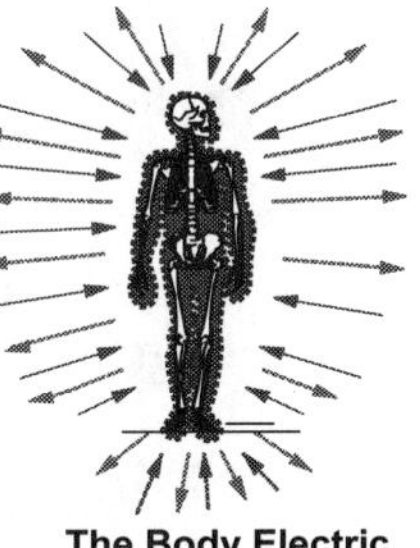

The Body Electric

Implications to therapeutic healing

Controlled cellular dedifferentiation and redifferentiation, and the cellular regeneration system

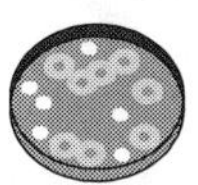

Bombshell: Amplified EM healing

- *Amplified disease anti-engine* quickly reverses disease.
- Unified field theory approach and supersystem analysis has broken the basic mechanism for healing itself.
  - The Sachs unified field theory applies and has been partially fitted.
  - Evans-Vigier O(3) electrodynamics can properly model it, including both the internal and external electrodynamics and effects of engines.
  - *Direct engineering development is now possible*.
- Experimental proof of results obtainable already exists in the French hard medical literature.
- A medical revolution is in the offing. Must change from century-old mindset against EM healing, and discover higher group symmetry EM, precursor engines and anti-engines as used by living biological systems.

Potential Medical Applications

- Non-traumatic cancer treatment
  - Cure cancers and leukemias of all types
  - Remove pre-cancerous state; *prevent promotion.*
  - Restore damaged system

- Treat and cure infectious diseases
- Cure AIDS and restore immune systems
- Defense against biological warfare
  - Quickly cure virulent diseases from unknown agents
  - Treat, cure BW victims after symptom developed
  - *Both civil populace and armed forces*
- Dramatically lower health costs
- Rejuvenate the aged

An observation for the future

- When the science of precursor engines is developed, physicians will be able to tailor these engine combinations to restore severed human spines, regrow severed limbs,* etc.
- They will also be able to direct the body's "normal growth" to reduce and direct birth defects, genetic defects, mental defects, etc.
- *Aging can and will also be reversed.*

* Such capability has been partially shown in lab animals by Becker and others. E.g., see R. O. Becker and Joseph A. Spadaro, "Electrical stimulation of partial limb regeneration in mammals," Bull. N.Y. Acad. Med. Second Series, 48(4), May 1972, p. 627-64.

A very important fact

- The two engineers who worked directly with Prioré, and who built his successful treatment machines, are still alive.
- A well-funded research program with these engineers could reproduce working machines and again demonstrate revolutionary results achieved by Prioré, Pautrizel, etc.
- Once the results are reproduced, this shows the machines do initiate revolutionary healing.
- The government scientific community could then have a very rapid Manhattan-type program to develop portable EM treatment machines, capable of saving millions of U.S. lives in coming terrorist BW attacks.

To treat and save mass casualties: Portable unit urgently needed (1)

- A small, portable, relatively inexpensive treatment unit, not requiring plasmas etc.
- Treatment in 2 to 5 minutes rather than 2-hours. Simple treatment application, almost anywhere.
- Effective in advanced stages after symptoms emerged, as well as prior to emergence of symptoms.
- Effective against a wide array of illnesses.
- Development difficult; use is easy. High school student can be trained to use it in 30 minutes.

To treat and save mass casualties: Portable unit urgently needed (2)

- Unsuccessfully urged crash program to develop, flood down units through emergency organizations in great quantities.
- Obviously is total anathema to the entire Big Pharmaceutical "drug 'em with expensive drugs" cartel and community.
- Once *precursor anti-engine treatment* is out of the genie's bottle, medical science will be dramatically altered.
- Could save millions of those future Triage Category 4 (black) patients who presently would be allowed to die.

Proposal to DoD, USAF, NIH, CDC, etc. in 1998

- In 1998 we proposed that U.S. government undertake crash Manhattan-style project to develop small, portable EM precursor engine treatment units for mass casualties
  - Extension of proven Prioré mechanism.
  - Develop in 2 years, at a cost of $40-100 million.
- First generation devices could save 70% of mass casualties. Second generation devices could save perhaps 90%.
- Patient treatment time 5 minutes. Produce thousands, flooded down to hospitals, fire stations, police stations, emergency crews.

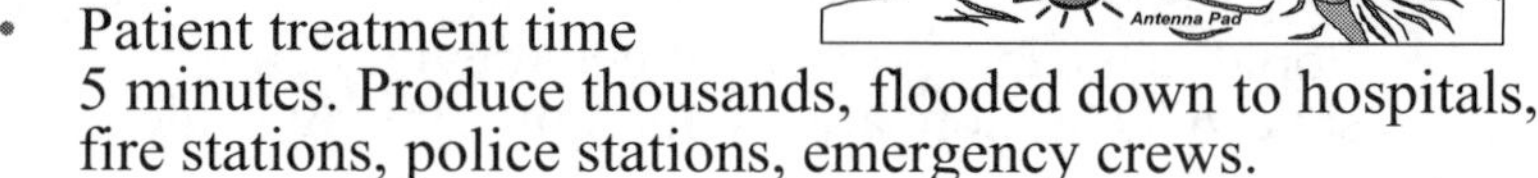

- 30 min. training to learn to set controls, etc.

Portable unit for rapid, effective treatment of mass BW casualties

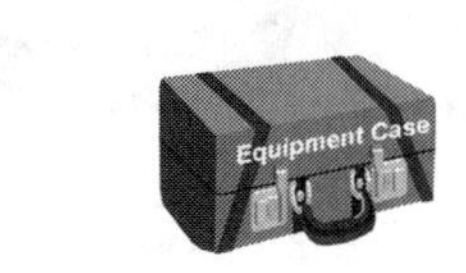

- Two blankets have antennas in them.
- Uses an advanced but simple concept.
- Eliminates plasma tube mixing etc.

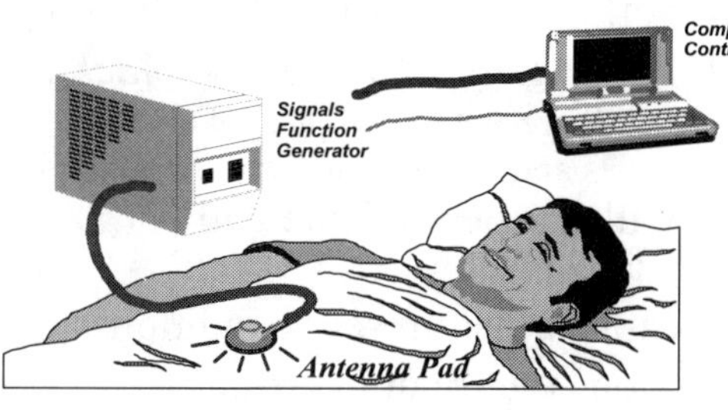

- Computer-controlled complex signal structure.
- Irradiation for 2 minutes.
- Three such treatments one week apart.
- Unit can be made self-powering on vacuum energy.

Anthrax spray attack (1)

- OTA study: Some 200 pounds released in a spray, on calm night over Washington, D.C. could result in from 1 to 3 million casualties. Three times that, if immune systems spread by QP weaponry
- Symptoms usually in 2-10 days
  - Flu-like symptoms; then 1-3 days later, shock and breathing problems cause death in up to 100%
- Antibiotics given early (before symptoms) can prevent infected persons from becoming ill
- Limited vaccine, reserved for military
- Portable bioenergetics treatment device could save 70% or more after symptoms already evidenced

---

* Multiply this number by at least threefold if immune systems are spread by QP.

Anthrax spray attack (2)

- TWO MEN, 1 SMALL A/C
- 100 KILOS ANTHRAX
- SMALL AGRICULTURAL SPRAYER
- FEW HOURS SPRAYING

- Highly lethal when spores inhaled.
- No present effective treatment at all unless treatment begun immediately.
- After symptoms appear, little or no chance to survive regardless of present treatment or not.
- *Most casualties will fall in Triage category 4 (black)*.
  - Useless to treat them with present techniques: likely die.
  - Must conserve medicines and vaccines in very short supply.
- Category 4 patients will just be set aside without treatment, and simply allowed to die.
- Millions of certain deaths under present conditions.
- NIH and other institutions totally uninterested.

Smallpox: highly contagious, no known cure, little vaccine*

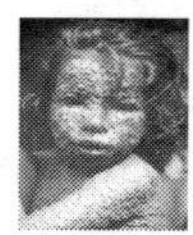

- Killed more than 500 million people in 20th century before eradicated in 1977; vaccinations ceased in 1980. (*A few million doses of vaccine left.)
- Russians kept pathogen, developed new and virulent strains, sold to terrorists.
- Symptoms appear in about 12 days: fever, headache, nausea. Rash similar to chickenpox appears, turns into hard blisters.
- *Highly contagious*, kills one-third of its victims. If immune systems spread, will spread easier and faster and kill two-thirds of its victims.
- If smallpox unleashed in any major city on earth, will eventually kill 2 billion persons, or about 1/3 of the human species — unless radical new cure is found and utilized.
- Could be contained by portable bioenergetics.

Urgent U.S. need for portable bioenergetics treatment machine*

- Five nations have developed quantum potential weapons using Bohm's hidden variable theory of quantum mechanics.
- Can induce disease engines in matter or bodies at a distance.
- Have developed "spreading" of immune systems in a targeted populace (cleverly tested in Gulf War against U.S. troops).
  - Cocktail mix of disease engines for 2 dozen different diseases.
  - Targeted area's human immune systems react to these engines when they are still in virtual state, just below quantum threshold.
  - Spread immune system's resources "thinned" across response to two dozen shadow diseases at once.

- Then hit with a terrorist BW attack; spread immune systems are overwhelmed easily and quickly.
  - Yield (casualties) of the attack is increased by factor of 3-5.
  - Some spreading of U.S. immune systems accomplished and is being maintained for reimplementation.

---

* Considering secret involvement of clandestine superweapons

Additional advantages

- Can be developed in a crash program in 2-3 years.
- After sunk costs, projected cost per unit in mass production is less than $150,000.
- Can be produced by the thousands.
- Large numbers of emergency operators easily produced.
- Emergency training could be by watching a 30 minute film, then performing one supervised group practice run.
- First generation equipment can save some 70% of the coming mass casualties from anthrax etc.
- Basis proven in France in thousands of lab animal experiments, documented in scientific literature.

Outcome of 1998 proposal

- No one had any notion what we were talking about, nor did they care one whit.
    - At NIH, we never got out of the "policy" section.
    - Seeing Gulf War Syndrome, NIH shipped package to DoD.
    - Uninterested in a scientifically proven method for revolutionary cure of cancer and infectious diseases.
- Apparently no one checked a single French scientific reference. Not a single scientist called to discuss the mechanisms, quality of the proofs, French literature, etc. Absolutely no interest whatsoever.
- But exercises and studies show we are in horrible shape as far as effectively treating mass casualties.
- We are actually much worse off than the studies show, due to QP augmentation by immune system spreading, EM BW, and carrier robots.

Weapons applications (already developed or in development)

- EM engines for any disease or disability
- Delivered by normal EM signals, QP weapons, robots, carrier robots.
- Attack entire populace or any fraction desired.
- Induce diseases desired.
- "Spread" targeted immune systems so BW attacks by international terrorists have extraordinary success.
- Render vaccines etc. ineffective.
- Cripple or destroy a populace slowly or with great speed, as desired.

Shadow disease state and immune system reaction

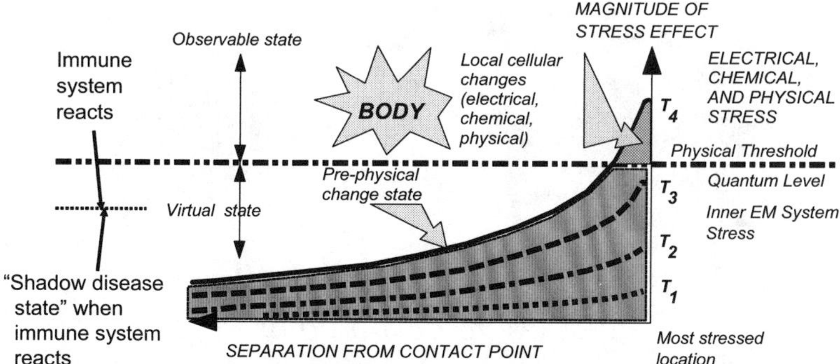

- The *shadow disease state* is that structured *virtual state* level of the promoting disease engine template, at which the immune system reacts before disease is observable.
- Eventually one will have a yearly exam, and then a quick treatment to remove all shadow state conditions prior to actual disease development. Will include aging rehabilitation.

Spreading (thinning) the immune system's ability to meet a new pathogen

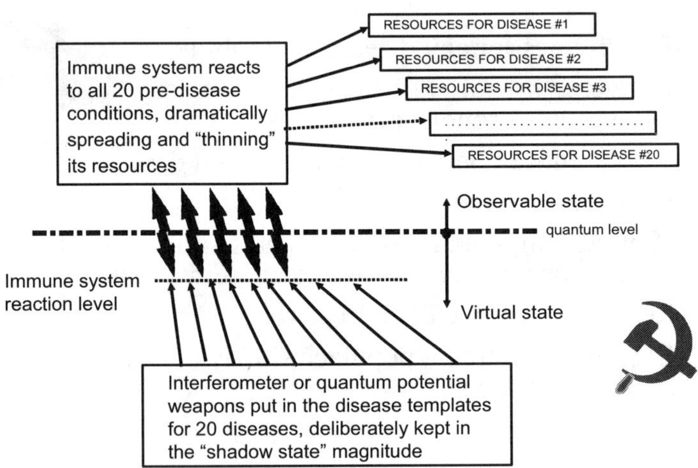

Gulf War Disease: Finding and statement*

"...those who served in the Gulf War are ill and their illnesses go beyond that which is explained by stress or psychiatric diagnosis."

---

* Research Advisory Committee on Gulf War Veterans Illnesses, 2002

Gulf War Disease (1)

- KGB test of closely controlled immune system spreading by QP weapon with "cocktail mix" in shadow state.
  - Americans, British, Canadian (ABC) troops were given harsh mix of inoculations; French troops (and native populace) were not.
  - ABC immune systems were temporarily depressed for some months by the inoculations, but not French (or native populace) immune systems.
- Shadow state effects in innoculation-lowered ABC troop immune systems much more pronounced than in French and natives.

Gulf War Disease (2)

- For vaccinated ABC troops, Russians thus could lower magnitude of shadow state QP "cocktail mix" necessary to produce a reactionable effect.*
  - Would still affect ABC troops' immune systems
  - Would not affect French and native immune systems.
- Using this test with dramatically different effects in ABC troops versus French troops was a probe to see if the U.S. recognized what had happened.
- U.S. scientists and intelligence analysts did not recognize it at all, and still do not understand it.

---

* The reader is reminded that charged particles negentropically order and integrate coherent virtual state energy patterns (precursor engines) to observable state EM force-field energy patterns (engines).

Gulf War Disease (3)

- Lowered ABC immune systems still form a statistical distribution.
    - KGB further lowered magnitude of shadow state cocktail mix, so only a *percentage* of ABC immune systems would be permanently affected.
    - With 45 years (9 generations of weaponization) experience in disease induction, this was simple for Russian bioenergetics scientists.
- A percentage of ABC troops' immune systems were spread and disabled; no effect on French and natives.
- Opportunistic diseases such as mycoplasma did the rest.
- Veterans who became ill because of their spread immune systems developed "cocktails" of diseases.

Gulf War Disease (4)

- Gulf War Disease thus is a cocktail mix of real opportunistic diseases; several such as mycoplasma being prominent.
- French troops and natives were essentially unaffected.
- A deliberate probe, so Russians could see if U.S. understood QP weapons and immune system spreading. They didn't and they don't. NIH doesn't want to hear it.
- See our expose on exchange of EM radiation between environment and deep interior of human bodies.
- QP exposure <u>*charged*</u> the veterans' bodies with the cocktail mix engines. Exchange of body fluids, or prolonged exposure to body emanations, could induce symptoms and diseases in some wives and children.

Indications of testing immune system spreading

- Subsequently, immune system spreading was tested in entire U.S. very carefully and at low level.
- Various diseases such as necrotizing fasciitis used.
- Signature is an absence of disease vector "path" record, but simultaneous *statistical distribution* of cases appearing throughout entire U.S.
- Again, a probe to see if NIH and the U.S. had their heads out of the sand. They didn't, and they don't.
- Some diseases such as mad cow disease in Canada and Europe almost certainly have KGB "steering" and "augmentation" effects. Avian flu or something similar could be the next test.

Conclusions

- Over several decades a strange strategic set of "combination" BW attacks developed and prepared.
- BW strikes on the U.S. by any group, will likely be augmented by FSB/KGB using QP spreading of U.S. immune systems.
- Can gradually raise the damage with other diseases such as mad cow disease, SARS, avian influenza, mycoplasma pneumonia, etc. Or, e.g., with normal flu, for epidemic.
- Adaptable for rapid massive destruction or gradually escalating destruction. Destroys the U.S. either way.
- Combinations provide an "over-the-threshold" capability to insure full strategic destruction of the U.S.
- NIH and U.S. government unaware.

(SLIDE BRIEFING CONTINUED)

PART F: Psychoenergetics and Its Weaponization

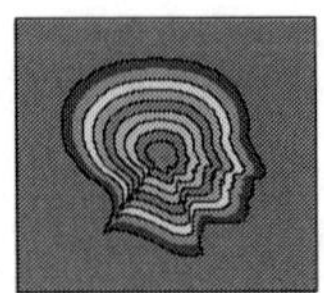

Psychoenergetics and its weaponization

Use of infolded precursor EM engines against mind and mind operations

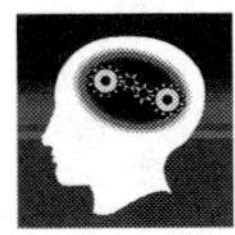

Section contents

- **Intelligence statement.**
- **Basic psychoenergetics.**
- **Model of living system.**
- **Mind levels and mind/body coupling.**
- **LIDA machine and tests.**
- **Personality mechanism.**
- **Model of life and death.**
- **Biogenesis.**
- **Captains Button, Svoboda, and Hess.**
- **Present mind control and new psychoenergetics.**

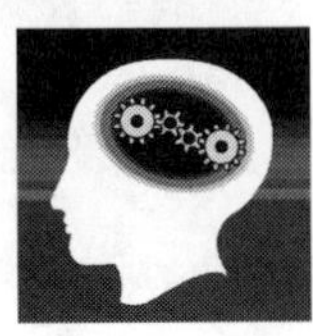

U.S. Intelligence statement on Soviet Psi research*

- Sooncr or later, Soviets may be able to:
  - Know contents of documents
  - Know troop and ship movements
  - Know installation locations and nature
  - Mold thoughts at a distance
  - Disable equipment at a distance
  - Kill persons at a distance
- **Comment**: These capabilities *and more* exist now
- Soviet psychoenergetics is a *physics*, not Western parapsychology as commonly believed

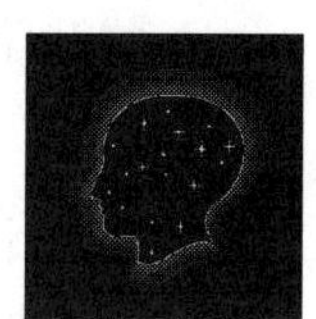

---

* Controlled Offensive Behavior - USSR (U). DIA Report ST-CS-01-169-72, July 1972.

A statement necessary for Western culture and scientists

- When God created life and living systems, He also created the *process* for making life and a living biological system.
- The process is still there and it exists as a *process in and of this 4-spatial world*.
- Once the process is discovered and modeled, technology can be developed to apply the process at will.
- So scientists can scientifically study the process and *develop the technology to use it*.
- Can make a living mind and living biological system, and tailor it to form and whatever behavior is desired.
- That superweapon science is called *psychoenergetics.*

A major Western problem (1):

Deep cultural bias against engineering the creation of life as a special electrodynamics

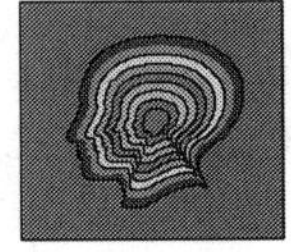

- Debate endlessly, deride experiments, etc.
- No definitions for intelligence, consciousness, life, living system.
- Do not understand mind dynamics as special EM dynamics on the time axis, using time-polarized photons and EM waves from quantum field theory.
- Have many viewpoints of living mind/body system; most favor materialism's "meat computer" notion.
- Very fuzzy on unconscious vs. conscious, and interrelation of the two.
- Do not understand "unconscious" mind is *multiply conscious*. A half century behind FSB/KGB psychoenergetics scientists.

A major Western problem (2):

Deep cultural bias against engineering the creation of life as a special electrodynamics

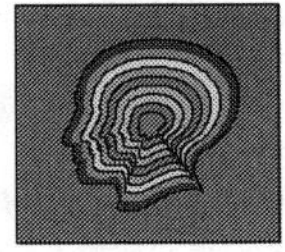

- Notions rather 3-spatial and mechanical. No notion of:
  - Mind operations as electrodynamics on the time axis (time-polarized photons)
  - "Engines" and "precursor engines" concepts.
- No apparent use of higher group symmetry electrodynamics in biology; but just standard EM.
- Some good experiments and projects nonetheless. Aleksandr has proposed five axioms of consciousness, which is at least a decent beginning.
- Sometimes must state to Western scientists: *"When God created life, He created the process for creating it. That process exists and can be utilized."*

Choice of fundamental units is arbitrary (1)

- The choice of fundamental units in one's physics model is completely arbitrary.*
- A viable physics model can be made using only a single fundamental unit, and it has been done.
- E.g., choose the joule (energy) as the only fundamental unit.
- All other "fundamental units" are functions of energy alone.
- Time itself, e.g., becomes a pure function of energy.
- Hence can speak of "extracting energy from time".

* E.g., see discussions by Jackson in his Classical Electrodynamics, 2nd and 3rd editions

Choice of fundamental units is arbitrary (2)

- In a model using the joule as the only fundamental variable, time itself becomes a function of energy.
- One function of energy can be changed into another. Energy can neither be created nor destroyed, but only changed in form.
- *Time* can be changed into energy and vice versa. And so on. *Space* (L^3) also is purely a function of energy, and energy can be extracted from it.
- Time has correlates to any energy entity, such as mass-energy, momentum, "body", field, photon, etc.
- Since EM energy occurs in flows which can be engineered, everything becomes *engineerable*.

Forefront Western view

- Igor Aleksander, emeritus professor, senior investigator in neural systems at Imperial College, London:
- *"When it comes to consciousness nobody really understands what they are talking about."*
- Some researchers advocate consciousness is an illusion.
- Strident dispute occurs over what is precisely meant by "machine intelligence". Can machines develop intelligence, or are they clever programs that only "mimic"?
- Others insist humans are only clever machines; brain is a "meat computer" with the mind itself just an illusion.*

---

* Aleksander predicts that in the future machines will themselves at least claim to be conscious. In his view, the machines are already beginning to "awaken". Some others in the forefront of mind and consciousness research share his view

Basic psychoenergetics

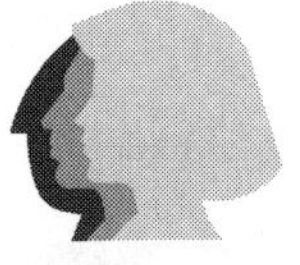

- <u>*Mind*</u>: "Sophisticated EM engine" in the time domain.*
- <u>*Mind operations*</u> on time-axis have coherent virtual-state projections into 3-space body (brain). Charges integrate.
- Integration of coherent virtual 3-space changes, correlated to mind's intent, forms coherent, real 3-space EM inputs to brain and nervous system.
- Movements of body elements give reverse rotational projections from 3-space to mind.
  - Coherently integrated to perceivable state to give feedback and "personal identity".
  - Engineerable with higher group symmetry EM.

---

* There are four photons in quantum field theory: Two transverse, one longitudinal, and one time-polarized. So there do exist photons (and therefore electromagnetics) active on the time domain. Combined longitudinal and time-polarized photons <u>*observed*</u> as a spike of scalar potential (i.e., as a voltage spike).

New view of a living biological system (1)

- The mind and its operations is an engine in time domain.
- The body and its force functions is an engine in 3-space.
- The mind-body linkage mechanism is an engine in pliant mass with rotations in 4-space. Thus the "triad system".
- All three engines can be engineered at will with higher group symmetry electrodynamics. Each has its precursor.
- The engines can be individually made, recorded, stored, joined, activated, etc. to form real living biosystems.
- Living robots can be formed (including at a distance) by insertion of triad precursor systems in pliant matter (electron gas, plasma, Whittaker interior of EM signals).

New view of a living biological system (2)

- The "life-forming process" given by the Creator is still present and available!
- Any type of living system is engineerable.
- One can make a toaster, a napkin — any bit of mass — alive. Some specialists already feel that "machines are awakening".*

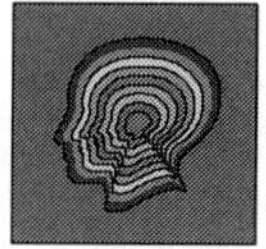

- Energetics scientists have made living nonmaterial robots that can reside in any bit of matter and adapt its charges to perform the functions of its "body".

---

* For a thought-provoking article, see Igor Aleksander, "I, computer," NewScientist, 19 July, 2003, p. 40-43.

Delay between mind-body and body-mind coupling mechanisms (1)

- A <u>mind change</u> (dynamic) on the time-axis creates a virtual projection component in 3-space (in the associated body).
  - "Intent" involves coherent "continued" mind changes, producing continual coherent virtual projection components in the 3-space body.
  - Coherent serial summation (integration)* in the body produces delayed observable photon changes in the 3-space body, correlated to mental intent.
- Hence the coherent mind intent produces observable but slightly delayed body changes in nervous system.

---

* The coherent summation process is provided by charges q absorbing virtual changes, coherently integrating them to quantum level, and re-emitting observable photons. Hence psychoenergetics cannot be understood without the solution to the source charge problem, accomplished by the present author in 1999 and published in 2000 and subsequently.

Delay between mind-body and body-mind coupling mechanisms (2)

- A body-change in 3-space creates a virtual projection component on the time-axis, in the associated mind.
  - Coherent body changes produce coherent "virtual state" projection components on the time-axis, in the mind.
  - Coherent serial summation (integration) produces perceivable, delayed (time-polarized photon) changes in the mind, correlated to body responses.
  - Hence the coherent body responses produce perceivable but slightly delayed mind changes.
- The total mind-to-body-to-mind delay/feedback loop creates the "sense of self" and sense of "my body".

Velocity of an object involves rotation out of 3-space toward the time dimension

- Necessary in order to understand the mind-body coupling mechanism.
- Body dynamics initiate on the L axis; mind dynamics initiate on axis T.

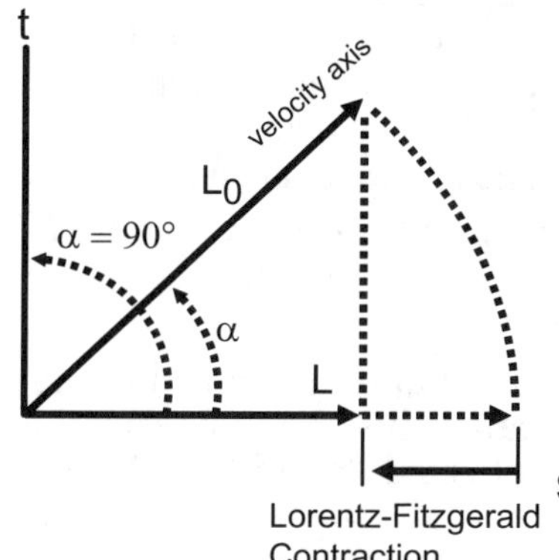

- An object moving with velocity $v < c$ along axis L has rotated from 3-space axis L toward the time dimension t by a small angle a.
- The object's "projection" back into 3-space, on axis L, is shorter, providing the *Lorentz-Fitzgerald contraction*.
- When velocity $v = c$, a full right-angle turn has been made, and the object has "lost" its projection along its direction of motion (axis L).
- Hence to the external observer it appears as a transverse 2-dimensional plane moving along L at the speed of light; in short, a *photon* or a *light wave*.

"Velocity" of a temporal change involves rotation out of the time axis into 3-space

- Necessary in order to understand the mind-body coupling mechanism.
- Mind dynamics initiate on the t axis; body dynamics initiate on axis L.

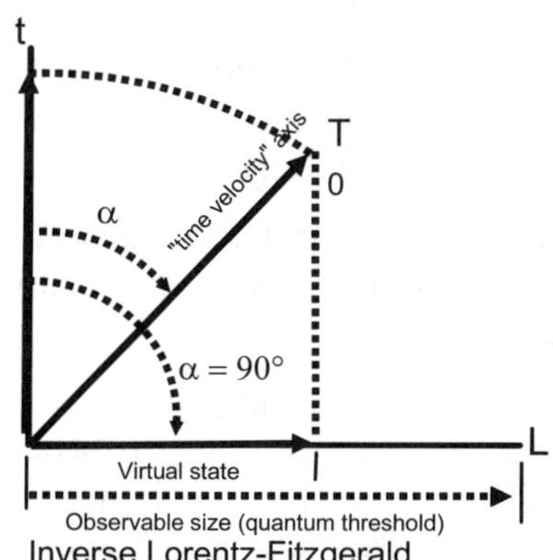

- Temporal photon's "velocity" (analogous to $v < c$) along axis T has rotated slightly away from time axis T toward 3-space axis L by small angle.
- Projection of the rotation, back into 3-space, on axis L, appears in the virtual state on axis L. Call this the *inverse Lorentz-Fitzgerald contraction*.
- When temporal velocity is analogous to $v = c$, a full right-angle rotation has been made, and the object has "lost" its projection along its temporal direction of motion (axis T).
- Integration on axis L of *coherent* serial changes of such temporal "mind" rotations produces real, observable photons on L, that are direct analogies of "mental intent".

How personality, awareness, and sense of being in the world are established (1)

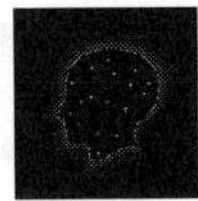

- Time-EM mental intent projects virtual EM changes into the mind-matter interface's 3-space body side.
  - Via charge negative entropy integration, continued coherent intent projections coherently integrate into observable EM input, after slight time delay.
  - Servomechanism theory applies in body's response.
- As body responds, physical response projects coherent virtual changes back through the mind-matter interface and into mind (the time) realm.
- Continued coherent response virtual projections integrate into "mentally observable" scalar photon input, after slight time delay. This completes the "mind-body" loop.
- Time delay in loop provides sense of self and conscious awareness of the body and of the physical world.

How personality, awareness, and sense of being in the world are established (2)

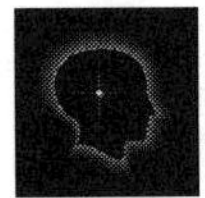

- Meanwhile, other physical universe change projections arrive in time domain which don't coherently integrate.
- <u>*Coherence to intent*</u> separates the "self-responses" of the body, from the "non-self-responses" of the external physical world which are <u>*incoherent to intent*</u>.
- The delay in the complete delay loop, and comparison of response to intent, creates the sense of self and living ("my" body, "my" mind).
- Comparison to the "self's" <u>*response physical world*</u> (the body) and the non-responsive or "external" physical world perceptions, creates the sense of the self being both "in" the physical world and also "separate" from it.

Mind-body coupling mechanism at work

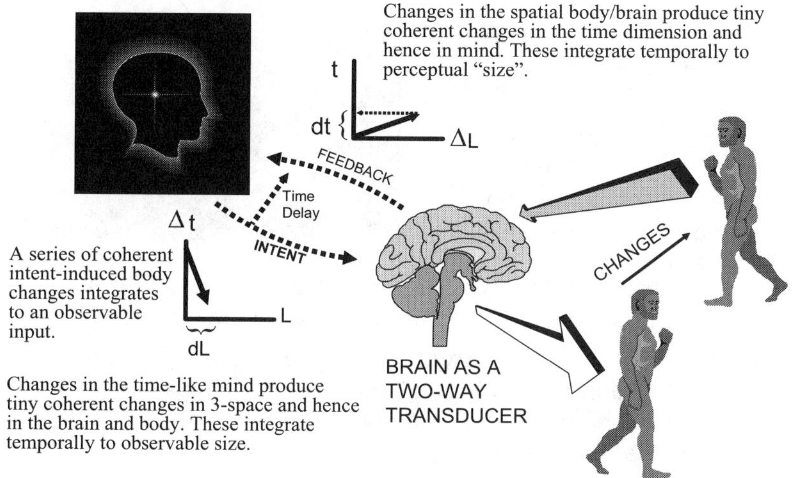

Mind levels and life and death (1)

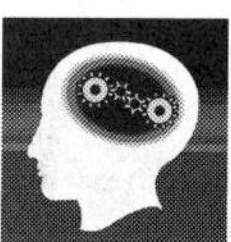

- The conscious mind is a *serial* processor.
- The unconscious mind is totally conscious, but *multiply* so. It is a *massively parallel* processor.
- The conscious mind is a "periscope" projected up from the unconscious (*up periscope!*) to deal with the "photon-interaction" perceived serial reality that constitutes the physical world.

Mind levels and life and death (2)

- In sleep, the conscious mind withdraws into the unconscious mind (*down periscope!*).
- In death, both the conscious mind and the unconscious mind separate from the body linkage permanently. The conscious mind is withdrawn into the unconscious mind permanently, on the time axis.
- Hence the "dead" individual is in very much like a deep dream state. Psychologically, one experiences the result of what one has "filed" and done. *What one has sown, now one also reaps.*

Levels of consciousness and unconsciousness

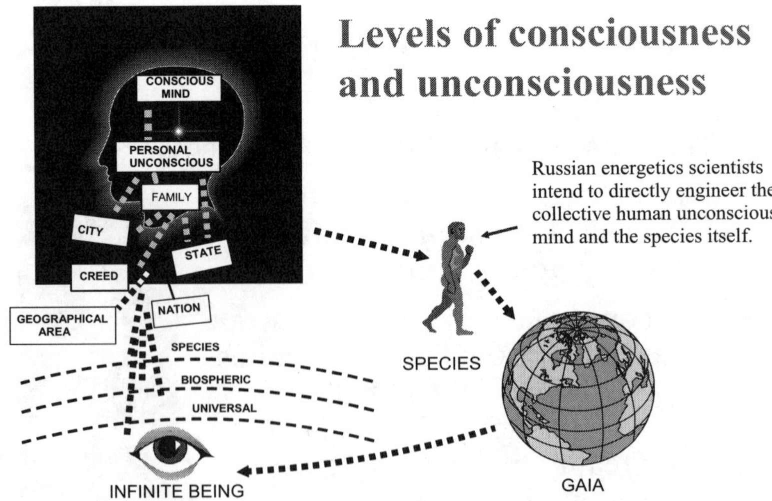

Mind levels are subject to engineering, attack, control, change, and destruction by higher group symmetry electromagnetic (psychoenergetic) technology.

A living biological system (psychoenergetics view)

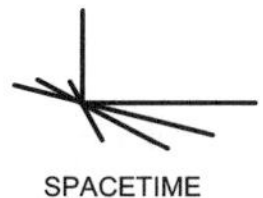

The body's cells and their parts are linked by a mutual quantum potential. Mind and thought (from a mindworld) occupy the internal SW biwave structure of the QP to that body, being continually disconnected and reconnected by the photon interaction generatrix for the flow of time.

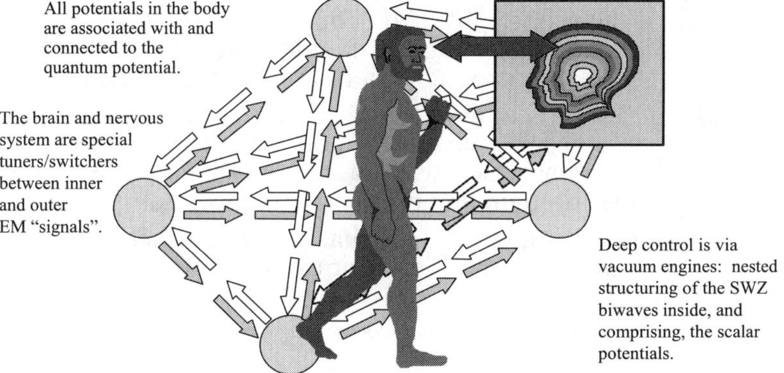

Mind levels and life and death (3)

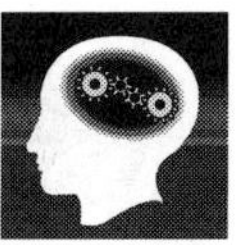

- Each "dead" mind still exists in its unconscious, which further exists in the collective human species unconscious on various levels — all in time domain.
- Death and life are therefore directly connectable (both exist in the same time)
- Dead (disconnected) minds can be engineered (in the time domain) as well as living minds connected to functional physical matter "bodies".
- A mind-engine on the time domain, without a connected body-engine in 3-space, is what the "death state" is. However, the conscious mind portion is withdrawn into the unconscious portion.
- The collective human species unconscious can in theory be "engineered", as can the biospheric collective unconscious for all species, etc.

Mind levels and life and death (4)

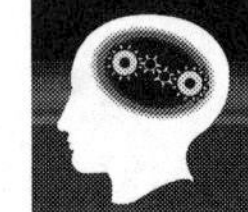

- Present biogenesis experiments occur in a nonliving disconnected *mindworld environment* as well as in a living *mind-connected-to-body environment*.*
- Produce the proper physical "body" tuner (mind-body coherent coupling) and body engine, and a "dead" mind reconnects, producing a living biosystem again.
- "Charge" and its integration of virtual state 3-space energy into observable correlated photons in 3-space provides the basic two-way intercommunication between mind (time) domain and physical (3-space) domain.
- Manipulating coherent L^3 virtual energy entities is manipulating "observable" entities on the time-axis.

---

* I.e., it follows since the experiments occur in time as well as space.

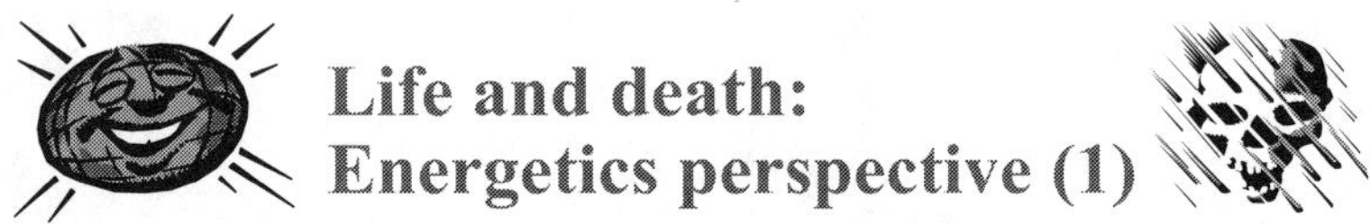

Life and death: Energetics perspective (1)

- A mind engine, coupled to a mind-body link engine and a body engine functioning in matter, is a living organism.
- Mind (time-engine) in collective species unconscious but separated from a functioning mind-body link and body engine in matter, is *L^3-dead* (separated from body) but still electrodynamically active in the time domain.
  - Still "living" (existing and functioning) in the time-domain, it exists in the time domain as time-polarized EM dynamics.
  - The "dead" or bodiless functioning mind can be *recoupled* to a body-engine, by a suitable mind-body coupling engine.
- Thus one can directly engineer a living biological entity, including building its mind-engine, its body-engine, and its mind-body coupling engine.

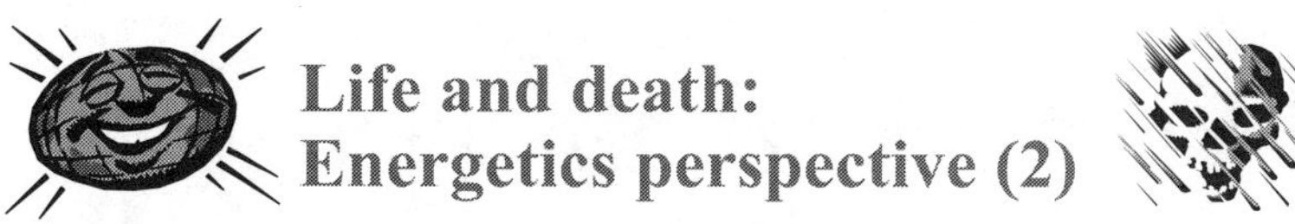

Life and death: Energetics perspective (2)

- All mind engines from every living system ever in the universe, pervade all spacetime since potentials do and did so and they reside in the 4-potentials. Inner structuring of EM is a vast world having so-called "Akashic records."
- We are immersed in a completely "living though dead" universe as far as mind engines are concerned. The entire physical 4-universe is *L^4 -alive* but much is also *L^3-dead*.
- Mind engines can be engineered and constructed, as well as spatial body engines and mind-body coupling engines. They can be connected or reconnected to again form living beings (special biological systems).
- Reconnecting an L^3-dead mind to a suitable L^3 body is *physical reincarnation* of an L^4 biological organism.

Selected psychoenergetics tests

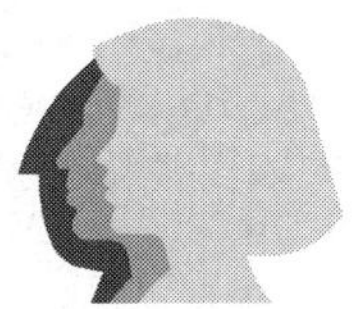

- LIDA machine
- Lisitsyn decomposition
- Captain Button
  - Points demonstrated
- Captain Svoboda
  - Points demonstrated
- Captain Hess
  - Points demonstrated

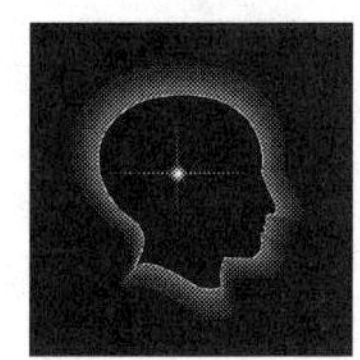

Russian LIDA Machine*

- U.S. Patent #3,773,049.
- For treatment of neuropsychic and somatic disorders, such as neuroses, psychoses, insomnia, hypertension, stammering, bronchial asthma, and asthenic and reactive disturbances.
- Pulsed RF field, pulsed light, and pulsed heat.
- Positive therapeutic results in 80% of patients.
- Early psychoenergetics device, quickly superceded, hence released for medical use.

* W. Ross Adey, "Biological effects of low energy electromagnetic fields," Proc. NATO Adv. Study Inst. on Advances in Biological Effects and Dosimetry of Low Energy Electromagnetic Fields, Plenum Press, New York, 1983. Study shows West knew and knows only ordinary electrodynamics and EM, and has no knowledge of real psychoenergetics. Not even aware that EM force fields do not appear in massfree spacetime, but only curvatures of spacetime appear there. Unaware that mass is a component of force, by $F \equiv \partial/\partial t(mv)$.

Typical Miller-Fox-Urey biogenesis experimental apparatus (1)

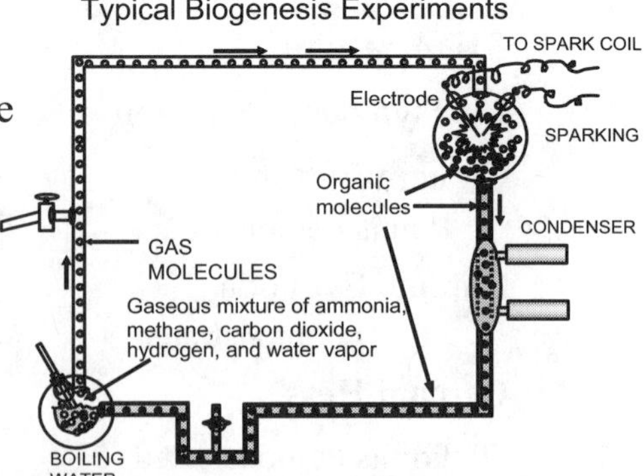

- Engines of and from all life forms ever on earth or in the universe exist in the ambient vacuum potential.
- The Miller-Fox-Urey experiments were not in a "sterile" environment.
- Heat (IR) and UV constitute (1) a harmonic interval with a difference frequency, and (2) amplifying by the process previously shown.

Typical Miller-Fox-Urey biogenesis experimental apparatus (2)

- These experiments "kindled" previously living forms from the available gases etc. by amplifying the ancient living engines from their virtual state to the observable state, so the masses were molded into the physical states of their templates.

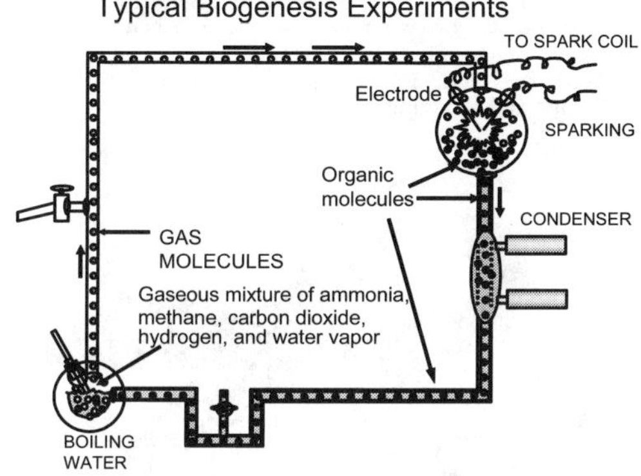

- KGB psychoenergetics weapons scientists understand this.

Three spectacular FSB/KGB psychoenergetics tests in U.S.

- Captain Button's bizarre flight to his death on 2 April 1997.
- Demonstrated control by switched persona.
- Target individual performing technical tasks.
- Captain Svoboda's "death dive" on 27 May 1997.
- Demonstrated switching a fractional conversion.
- Simply switched Svoboda's sense of up and down.
- Svoboda reacted with new fraction, diving to her death.
- Captain Hess's strange suicide on 4 March 1998.
- Demonstrated ability to function coolly in spite of autonomic nervous reflexes until body is nonfunctional.
- Example of complete conversion.

Captain Button's bizarre flight to his death on 2 April 1997

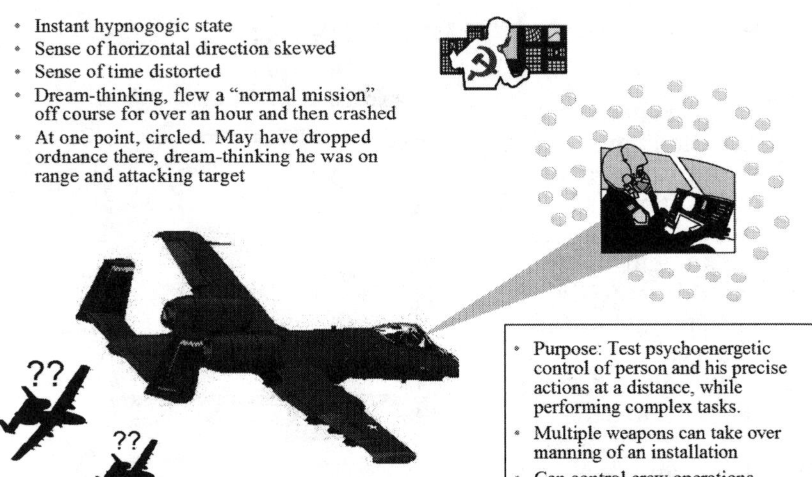

Captain Button's flyaway path to his eventual fatal crash

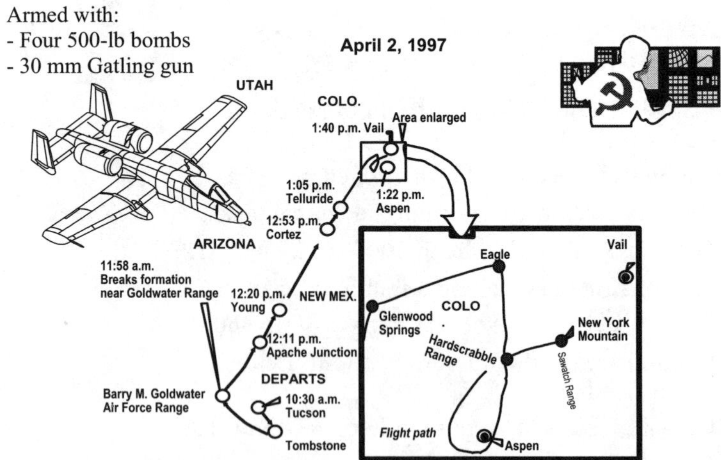

Many distinctive Russian tests have occurred in the month of April over the years, in preparation for the annual May Day report to the die-hard communist faction of the KGB.

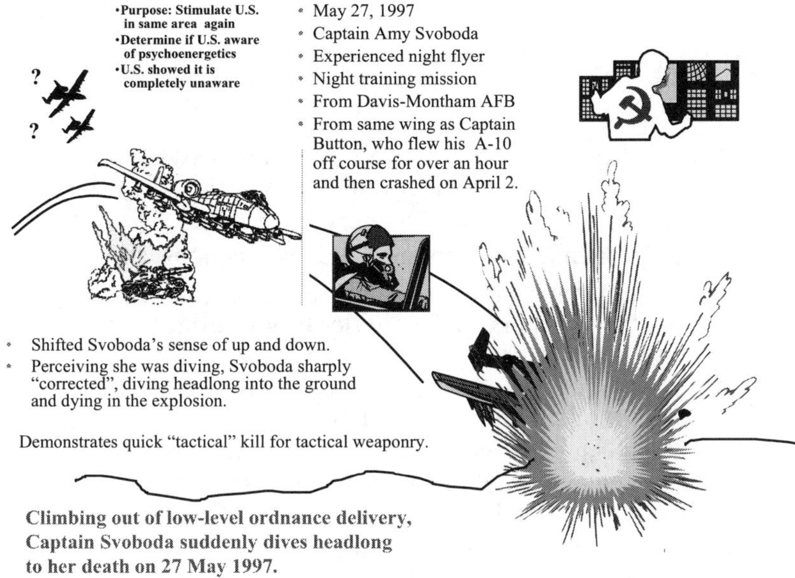

Captain Hess's death: conversion including autonomic nervous system

- Capt. Gordon Hess, NY Army National Guard.
- Early morning jog, Mar. 4, 1998, Ft. Knox, KY.
- Veered aside, sat down, and stabbed himself 26 times with relatively small Leatherneck utility knife.
  - Penetrated left heart ventricle twice; each a disabling blow.
  - Penetrated lung with hemorrhaging 4 times (disabling).
  - Penetrated liver once.
  - Four L-shaped wounds (calmly twisted the knife).
- Demonstrates deep control of autonomic nervous system; calmly ignored pain, hemorrhaging, etc. altogether.
- Officially declared a "suicide". Impossible.
- Probably an acceptance test of Russian psychoenergetics weapon improvement from 1997 (Button and Svoboda).
- Suspended autonomic system reflexes in the conversion.

When suicide is not suicide

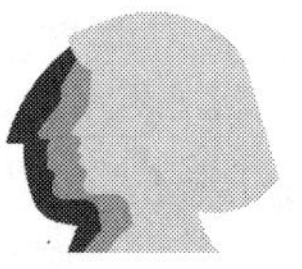

- When a fraction or all of one's persona is switched, then a different entity (from one's normal persona) is
  - In control of the body and nervous system
  - Ordering and carrying out the body's actions.
- Intentional "suicide" under such circumstances is not normal "single persona" suicide. It is murder because it is by direct intent and actions of an outside party.
  - It is still "intentional suicide" by the persona in control of the body at that time.
  - It is not suicide by the individual's normal persona.
  - Analogous to "deliberately switched schizophrenia", with intent of what the "second persona" will do.
- Psychiatrists, clinical psychologists, and law enforcement officers cannot properly assess it.

Present mind control capabilities

- Conversion or control mind directly.
- Complete control over mind and body, *including autonomic nervous system*.
- Consider controlled actions desired for a given purpose:
  - Each requires a persona, characteristics, emotions etc.
  - In short, one "conditioned" personality (engine).
  - Can have multiple such conditioned personality engines available, each for different purpose, task, and mission.
  - Switch personality engines in controlled person at will.
- Carrier robots can transport multiple engines, including energetic, bioenergetic, psychoenergetic. Can "infect" multiple targeted persons or persons of opportunity.
- Converted person is a multiple purpose weapon system.

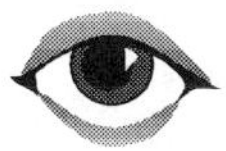

Psychoenergetics weapons teams mentally control all site personnel

Counters all sane mad* systems and nullifies human component of all weapon systems

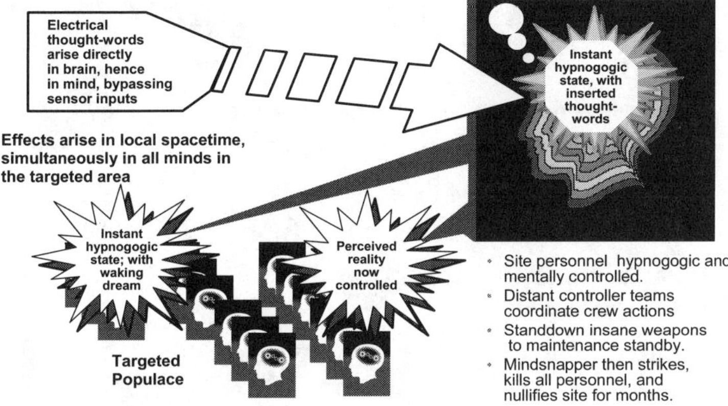

Indicators: Two A-10 incidents 1997, Captain Hess incident, LIDA device since 1950s.

---

* MAD = Mutual Assured Destruction.

Psychoenergetics weapons teams switch sleeper conversions in all crewmen

Counters all sane mad* systems and nullifies human component of all weapon systems

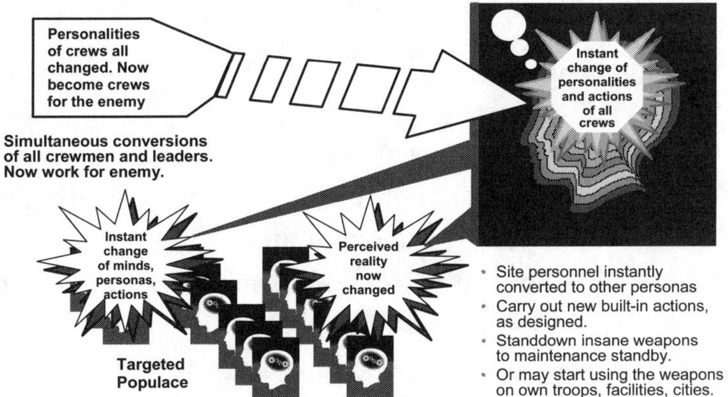

Indicators: Captain Button and Captain Hess incidents.

Propagation of Weapon Effects (1)

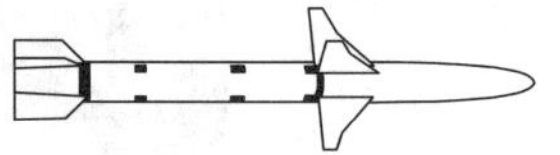

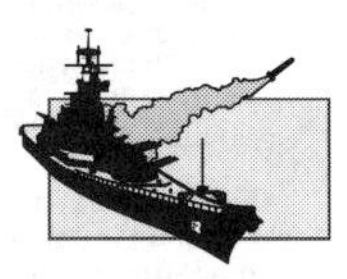

- *Conventional weapons propagate matter or energy or both through 3-space.*
- Matter (bombs, missiles, projectiles, etc.).
- Energy (EM force fields, sonic force fields, etc.)
- Basically one-to-one, or one "bus" to many-on-any.
- Finite propagation velocity v: where $0 < v \leq c$.
- Time delay: Shielding and pre-strike ` retaliation possible.

Propagation of Weapon Effects (2)

Four objects in multiply-connected spacetime superpose

10,000 miles in conventional space

- Structured quantum potential propagates vacuum ordering (vacuum engines), instantly.
- Zero time delay (through multiply-connected spacetime).
- No propagation "through space", no "velocity" as such.
- Easily one-to-many, without limit.
- Dramatic energy amplifier (gates, amplifies flux ordering).
- Effects arise in local spacetime; no shielding possible.
- No range limit, since striking distance always zero.
- QP weapons obsolete all other weapons.

Conversion and fractional conversion (1)

- *Conversion* is insertion of an extra variant of complete mind-engine of an individual, in that individual. Two separate personalities reside in the individual's body.
  - Control switched from one to the other at will.
  - Alters the behavior of that individual at will.
  - Can insert and switch between multiple variants.
- *Fractional conversion* is the insertion of an extra variant of a *fraction* of the mind-engine of an individual, in that individual. The fraction is for a *given area of behavior*.
  - The fractions can be interchanged (switched) at will.
  - Alters specific behavior of individual in one area, at will.
  - Can insert and switch between multiple fractions.
- Can insert and switch between more than two.

Conversion and fractional conversion (2)

- Can pre-insert conversions and fractional conversions, then switch at will, or as frequently as desired.

  - Can insert personas or fractions in an entire populace or large fraction.
  - Can massively switch all of them at once.

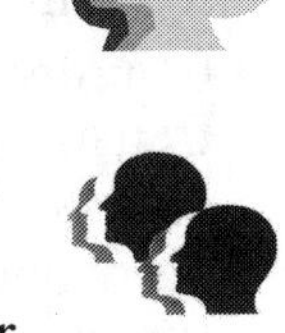

- Instantly changes the society and its behavior.
- With all coordination destroyed, utter chaos reigns.
- Massive conversions can cause a society to rapidly destroy itself, beginning very suddenly and everywhere in the society.

Sleeper conversions

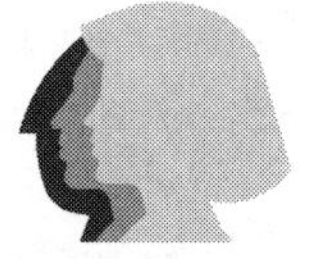

- Conversions or fractional conversions that are inserted, but not switched and activated for a lengthy period, are called “sleepers”.
- A few leaders, a fraction of the society, or all the individuals of a society can have sleepers inserted, ready for switching.
- Multiple personas or fractions can be in each individual, waiting for switching conversion.
- Tailored switching then allows “tailoring” of both the societal responses and the momentum of the converted society's changes and actions.

The new psychoenergetics

- Coupled with QP weapons using engines against mind and mind dynamics.
- Coupled with carrier robots each transporting hundreds of mind-control engine subrobots for unleashing.
- Now becomes a separate, great strategic threat more ominous than an all-out nuclear strike.
- Can be cleverly done so unsuspecting nation(s) never really know(s) what happened, till too late.
- Western science, military, and governments are unaware, unprepared, and nearly defenseless.

(SLIDE BRIEFING CONTINUED)

PART G: CONVERSION

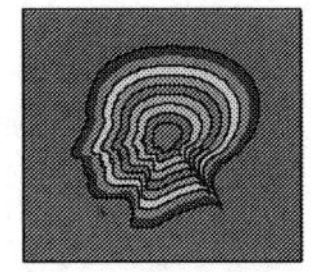

Conversion: Mind, behavior, and personality control as primary target

Unified energetics capabilities, including robots and engines

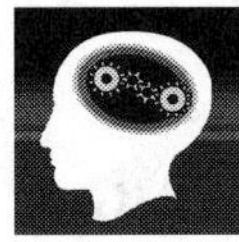

Section contents

- **Russian National Style**
- **Strategic attack plan(s)**
- **Tests and stimuli expected**
- **Advance indicators**
- **Defense and counters**
- **Conclusions**
- **Recommendations**

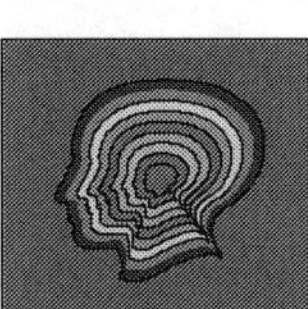

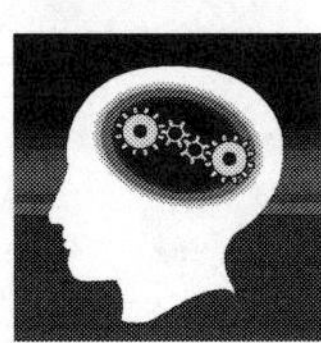

National style (Russian): Chessplayer approach

- Develop each weapon type (move) secretly.
- Always have elaborate deception plan.
- Multiple specific "stimuli" probes tested in advance of planned actual attack.

- Ascertains opponent unaware of "technique" and coming attack(s).
- Attacks lulled opponent unless countered.
- Continues rapid revision and improvement.
- When countered once, returns with improved techniques, new probing, and new attack(s).
- Never quits; always maneuvering to attack and win.

On elements of the strategic KGB attack plan (1)

- Only outline of probable type plan(s) seen:

- Attack to be strategic, massive (incredibly so).
- Number of robots/engines may be 100-200 million.
- Includes positive nullification of Israel, and of any U.S. retaliatory capability, in first ten minutes.
- May include nullification of Brazil and China also.
- Includes nullification of entire strategic strike capability of the United States — practical by the robots and QP weapons — in that 10 minutes.

On elements of the strategic KGB attack plan (2)

- Massive mind control conversions nullify both personnel, facilities, weapons, command and control, support facilities, populace.
  - Wholesale disruption and nullification of government officials, congress, military commands.
  - Total anarchy quickly.
- Defeat *prepared* in insertion/preparation phase.
- Insertion phase has been largely successful; U.S. to be defeated in operations phase just begun.

Expected tests and stimuli (1)

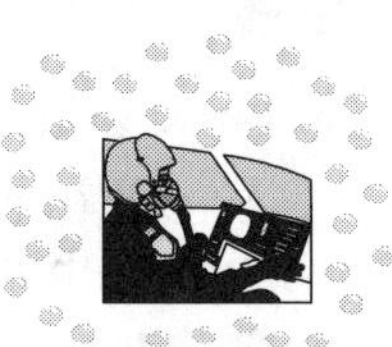

- Next generation engines and robots.
- Great new development beyond Captains Button, Svoboda, and Hess.
- Probing should be ongoing now, using conversions at selected levels.
  - One or two (few) high government officials or personnel surrounding or subordinates.
  - One or two personnel in critical facilities (e.g., nuclear).
  - One or two personnel in extremely sensitive intelligence agencies.
- Show multiple personas switching as probe to see if U.S. understands.
- Armada and extreme strategic attack planned for about a year later, after the exposures.

Expected tests and stimuli (2)

- Some tests should be bizarre.
  - Person performs some bizarre act or action, totally noncharacteristic.
  - Personas deliberately switched.
  - Psychiatrists treat it as “psychiatric problem”.
- Overwatch (via robots) assures that system does not understand, with 100% certainty.
- Tests should include such acts as:
  - Compromising national security.
  - Committing espionage but so it is easily detected.
  - Physical involvement such as with nuclear materials, biological warfare agents, drugs, etc.

Advance indicators expected (1)

- Signature indicators in major command and control facilities, sensitive agencies, critical weapons such as ICBMs and nuclear subs, strange symptoms in computers etc. Tests in Congress, Executive Branch, perhaps in judicial branch, and in JCS.
- Several high officials converted — at least two.
- Series of a few increasingly bizarre conversions.
- Perhaps a Supreme Court Justice or the Vice President, etc.
- Several high military officers converted.

- Conversions (few) in news media and public, slowly increasing.

Advance indicators expected (2)

- After some of the bizarre conversions, will then begin gradual "minor" conversions of public, public officials, military personnel, news media on broader scale.
- Opinions and media coverage gradually change.
- Similar opinion changes occur in Arab nations, Islamic populaces, Europeans, Africans, China, etc.
- Anti-Americanism steadily rises, "normal" terrorist incidents increase, incidents in U.S. increase.
- May be a special effect in the extremists of both left and right, racism, all social issues to foment conflict.
- Increasing conflict will start to unravel the entire society.
- UN should also be unraveled and effectively neutralized by strident differences arising and increasing.

Advance indicators expected (3): ***Orchestration***

- *Orchestration* of world arena actions by orchestration of leadership of factions, nations, and their actions.
- Opinions and media coverage gradually *orchestrated*.
- Changes occur in Arab nations, Islamic populaces, Europeans, Africans, China, etc., *centrally ordered*.
- Orchestrate special effects in the extremists of both left and right, racism, all social issues to foment conflict.
- Intensifying conflict starts to unravel targeted societies, control world nations' behavior and actions.
- UN, NATO, etc. to be unraveled and neutralized by strident differences arising and increasing.

Advance indicators expected (4)

- Conversions of foreign high officials, so that friendly nations begin to act as hostile nations.
- Strong, even bizarre developments against Israel.
- Another indicator will be terrorists — stimulated into unleashing major BW strikes etc., in our cities and populace — having unexpected effectiveness due to KGB QP *spreading of immune systems* in the U.S. populace.
  - Smallpox, anthrax high possibilities.
  - Ebola virus and necrotizing fasciitis, avian flu etc. also possible, as is smallpox.

Blurring of classifications

- Intelligence agencies are familiar with the definitions of "agent" vs. "agent of influence".
- Conversion and multiple personas gives switching of a given person between:
  - Normal person (one's normal personality).
  - *Agent of influence* (a second persona which is "oriented" in a specific direction, to take specific actions at its own volition), or
  - *Agent* (a third persona which actually communicates with the "agent handlers" etc. and works for them).
- In any case, the person and his personas are manipulated like a puppet on a string.

Defense and counter-measure capabilities

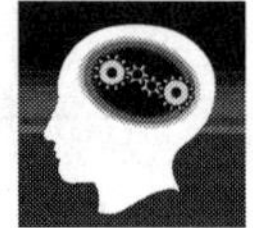

- Very few possible for U.S. at present.
- Urgent that knowledge of the capability and such hostile measures be spread.
- Israelis have some lesser countermeasures which must be rapidly expanded.
- U.S. must have immediate crash national program, with a target completion date of a year only.
- Must have strongest possible Presidential Directive, action, and total support from any agency as needed. Move now, or it will be too late.

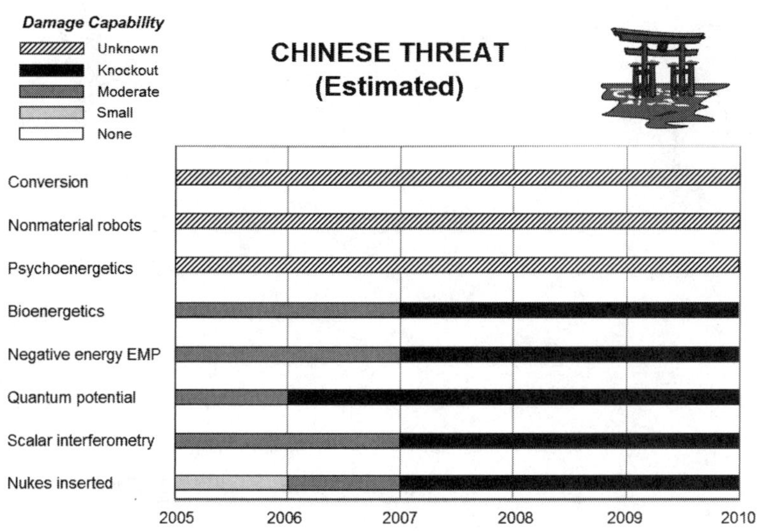
Damage Capability
Unknown
Knockout
Moderate
Small
None
CHINESE THREAT
(Estimated)
Conversion
Nonmaterial robots
Psychoenergetics
Bioenergetics
Negative energy EMP
Quantum potential
Scalar interferometry
Nukes inserted
2005
2006
2007
2008
2009
2010

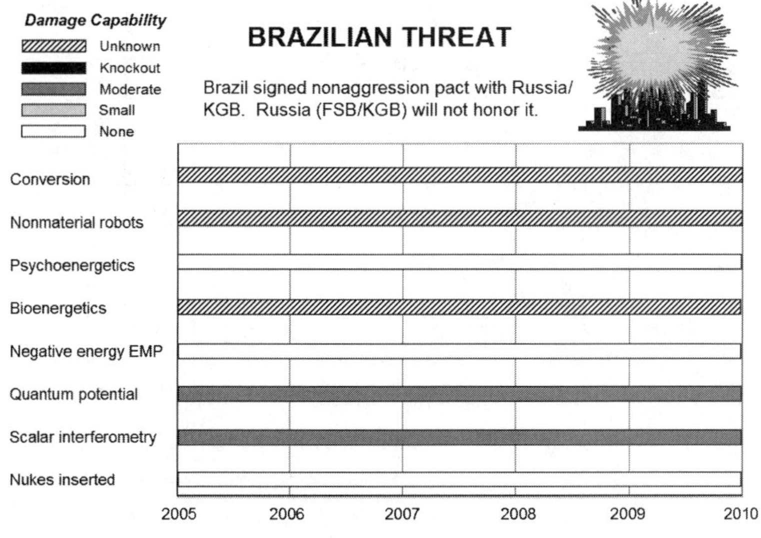
Damage Capability
Unknown
Knockout
Moderate
Small
None
BRAZILIAN THREAT
Brazil signed nonaggression pact with Russia/
KGB. Russia (FSB/KGB) will not honor it.
Conversion
Nonmaterial robots
Psychoenergetics
Bioenergetics
Negative energy EMP
Quantum potential
Scalar interferometry
Nukes inserted
2005
2006
2007
2008
2009
2010

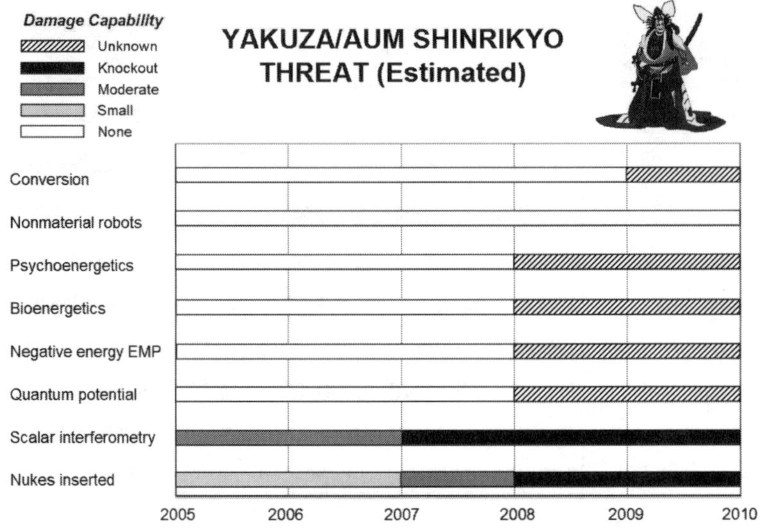

Damage Capability
Unknown
Knockout
Moderate
Small
None
YAKUZA/AUM SHINRIKYO
THREAT (Estimated)
Conversion
Nonmaterial robots
Psychoenergetics
Bioenergetics
Negative energy EMP
Quantum potential
Scalar interferometry
Nukes inserted
2005
2006
2007
2008
2009
2010

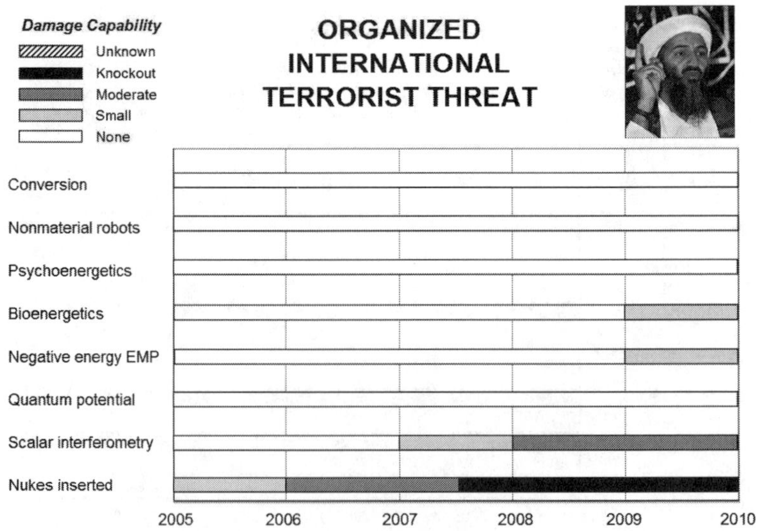

Damage Capability
Unknown
Knockout
Moderate
Small
None
ORGANIZED
INTERNATIONAL
TERRORIST THREAT
Conversion
Nonmaterial robots
Psychoenergetics
Bioenergetics
Negative energy EMP
Quantum potential
Scalar interferometry
Nukes inserted
2005
2006
2007
2008
2009
2010

Conclusions

- U.S. is in gravest peril in its history, as is humanity.
- Most stringent and fastest possible action is required.
- Two-thirds to three-quarters of human population may be destroyed; probably will be. *Eventually* the entire human species could conceivably self-destruct.
- Multiple strategic destruction threats already present, others converging.
  - Energetics weapons threat.
  - Bioenergetics weapons threat.
  - Psychoenergetics weapons threat, robots and conversions.
  - Terrorists weapons threat (nuclear and BW).
  - Chinese threat.
  - Yakuza/Aum Shinrikyo threat.
  - Orchestrated dissolution of U.S. society, Europe, UN, etc.

Recommendations (1)

- Crash nuclear scanning of all incoming shipping containers.

- Crash development of energetics science and some new counters in one year or less.
  - No more than two years available.
  - Immediate Presidential Decision Directive issued, highest priority.

- Crash development, mass production of fuel-free self-powering electric power systems taking their energy freely from the vacuum. *Essential*.
- Immediate cooperation with Israelis for initial counters.

Recommendations (2)

- Crash development of the science and new counters in one year or less. No more than two years available.
- Crash development of Prioré-type treatment devices to survive stimulated large terrorist BW attacks, with immune spreading and mass casualties certain. *The devices are absolutely essential*.
- Immediate "think tank" of most-capable young scientists to clarify the psychoenergetics problem and actions to be taken.

- Brief U.S. public on insertions of nuclear weapons and anthrax already accomplished, so will support the program.

Final comments (1)

- Less than 3 years before our total destruction.

- Any delay of prompt and drastic national action is fatal.
- Biggest problem is the U.S. scientific and intelligence communities, still sublimely resting in archaic EM theory and electrical engineering. The national labs are nearly useless for the correction effort, as are NSF, NAS, and NAE, DoE, etc.

- Leaders of the two communities must be bluntly shocked as drastically as Stalin shocked his Academy of Science in 1945.

Final comments (2)

???

- The President of the U.S. will have to deliver a "meat axe" ultimatum to the Intelligence Agencies, National Academy of Sciences, National Academy of Engineering, National Science Foundation, NIH, the great national laboratories, etc.

- Any science or intelligence leader opposing the program should be immediately relieved under full presidential wartime powers.
- It's either fish or cut bait immediately. *It's do or die*.
- For those of us who are religious, it is also a time for fervent prayers on behalf of our threatened nation.

(SLIDE BRIEFING CONTINUED)

PART H: APPENDIX; Problems in Western Science

Appendix 1: Problems in Western science

Energy science's flaws and limitations

Energetics: start by correcting the flawed CEM/EE model

- The classical EM and EE model is seriously flawed, and has been for more than a century.*
- Soviet energetics started by a massive correction and reworking (and extension) of CEM/EE.
- To understand energetics and energetics weapons, it is essential to first grasp the horrendous errors that are perpetuated in CEM/EE in our universities.
- More extensive systems of electromagnetics have been developed in physics (Yang-Mills theory, higher group symmetry EM, etc.) because CEM/EE does not adequately describe nature.
- The falsity of assuming force fields in space started with basic mechanics several hundred years ago, and was continued into CEM/EE by its founders.

* T.E. Bearden, "Errors and Omissions in the CEM/EE Model", at http://www.cheniere.org/techpapers/

Seriously flawed path of classical electrodynamics

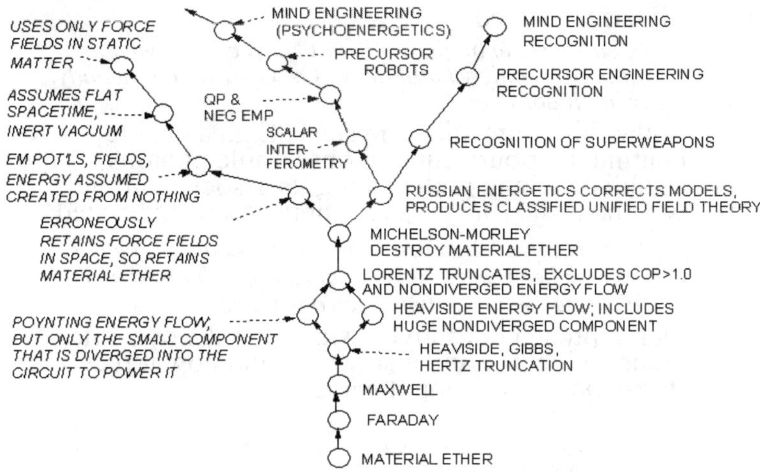

Electrodynamics and electrical engineering

- There is not now, and there never has been, an EE department, professor, or textbook that knows and teaches what really powers an EM circuit or system.
- Every EM circuit and system is powered by EM energy extracted from the vacuum by the asymmetry of opposite charges on the ends of the source dipole.*
    - Basis proven in particle physics since 1957.**
    - All EM field energy and EM potential energy in a circuit is extracted from the local vacuum by the source charges q of the circuit. None comes from mechanical energy input by cranking the shaft of the generator.

---

* See Steven Weinberg, Dreams of a Final Theory, Vintage Books, Random House, 1993, p. 109-110. Dipolarity includes the classical "isolated" source charge, when the charge's polarization of its surrounding vacuum is accounted. An isolated charge is actually an infinite bare charge in the middle, surrounded by an infinite virtual charge of opposite sign. The *difference* as seen by an external instrument is finite, and it is the textbook value of the *observed* "classical charge".

** See C. S. Wu et al., "Experimental Test of Parity Conservation in Beta Decay," Physical Review, Vol. 105, 1957, p. 1413. Lee and Yank received Nobel Prize

Nature of charge is unresolved

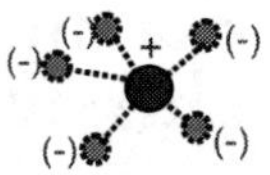

- Silverman stated:

  *"...curiously enough, we do not know exactly what charge is, only what it does. Or, equally significantly, what it does not do."**

- Some things are still a mystery. E.g., a charge continually pours out real observable photons, establishing and replenishing its associated field, with no *observable* energy input. As Sen** stated:

  "The connection between the field and its source has always been and still is the most difficult problem in classical and quantum electrodynamics."

- Our approach does give insight into the dynamic nature of charge and also where and how it receives its *nonobservable* input energy.

---

* M. P. Silverman, And Yet It Moves: Strange Systems and Subtle Questions in Physics, Cambridge University Press, 1993, p. 127.

** D. K. Sen, Fields and/or Particles, Academic Press, London and New York, 1968, p. viii.

The Source Charge Problem (1)

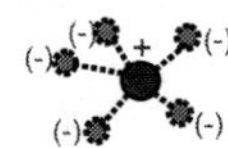

- Kozyakov** stated the problem more bluntly:

  "A generally acceptable, rigorous definition of radiation has not as yet been formulated. ...The recurring question has been: Why is it that an electric charge radiates but does not absorb light waves despite the fact that the Maxwell equations are invariant under time reversal?"

- We solved the problem in 1999 and published it.**

---

* B. P. Kosyakov, "Radiation in electrodynamics and in Yang-Mills theory," Sov. Phys. Usp. 35(2), Feb. 1992, p. 135, 141.

** (a) T. E. Bearden, "Giant Negentropy from the Common Dipole," Proc. Congr. 2000, St. Petersburg, Russia, Vol. 1, July 2000 , p. 86-98; (b) — "Extracting and Using Electromagnetic Energy from the Active Vacuum," in M. W. Evans (ed.), Modern Nonlinear Optics, Second Edition, 3 vols., Wiley, 2001; Vol. 2, p. 639-698; (c) — Energy from the Vacuum: Concepts and Principles, Cheniere Press, Santa Barbara, CA, 2002, (d) M. W. Evans, T. E. Bearden, and A. Labounsky, "The Most General Form of the Vector Potential in Electrodynamics," Found. Phys. Lett., 15(3), June 2002, p. 245-261.

The Source Charge Problem (2)

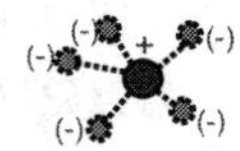

- First we state the problem:
  - Every charge continuously radiates real, *observable* EM energy, forming and replenishing its ordered fields and potentials.
  - There is no input of *observable* EM energy to the charge.
- Either the charge:
  - Creates energy out of nothing at all (falsifying the first law of thermodynamics), or
  - It receives, reorders, and integrates disordered *virtual state* energy from the vacuum into *ordered observable* energy (falsifying the second law of thermodynamics).
- The latter is true, and the second law requires revision. A charge continuously consumes virtual state entropy and produces observable state negative entropy.

The Source Charge Problem (3)

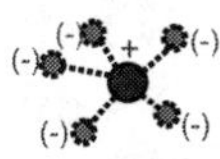

- There *is* continual:
  - *Disordered virtual photon energy input* to the charge.
  - *Reordering* of the input virtual energy, and
  - *Coherent integration* of the reordered virtual EM energy input into output as quanta (observable photons).
- So the charge continually radiates observable photons in all directions, extracting the necessary input energy from the vacuum subquantal fluctuations.
- We do not have to discover how to extract EM energy from the vacuum via a "new device"!
- Every charge in every circuit and system already does it, freely and continuously.
- We just must learn to properly catch and use the freely flowing EM energy from the vacuum.

EM energy occurs as sets of ongoing EM energy flows

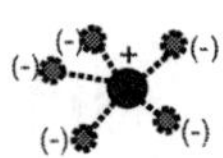

- Every EM field and potential — whether "static" or "dynamic" — is a set of ongoing EM energy flows.
  - Whittaker proved it for the scalar potential in 1903.
  - In 1904 he proved any field decomposes into differential functions on two scalar potentials.
  - If we further apply his 1903 decomposition to each of the "base" scalar potentials of the 1904 paper, EM field emerges as a set of energy flows also.
- Hence Whittaker's work also shows that — in producing its fields and potentials — the source charge is pouring out real EM energy continuously.
- *<u>Thus EM "static" fields and potentials are ongoing sets of EM energy flows, not "chunks" of static energy.</u>*

The source charge solution: Free EM energy from the vacuum

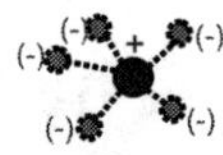

- The emitted observable EM radiation continually *<u>establishes and replenishes</u>* the associated spatial EM fields and potentials of the source charge, spreading at light speed in all directions continuously.
- There is no "static" EM field from a "static" source.
- The "static" EM field, its source charge, and the vacuum input form a nonequilibrium steady state (NESS) system of continual, observable photon emissions and flow from the seething vacuum.
- So is the "static" EM potential.
- Every EM field, EM potential, and joule of observable EM energy is already *<u>a set of ongoing free flows of EM energy</u>* extracted from the vacuum.

A charge is a dipolar ensemble (1)

- An "isolated charge" polarizes its vacuum medium.*
- Considered from a quantum field theory view, the isolated classical charge is a bare infinite charge in the middle surrounded by an infinite charge of virtual particles, of opposite sign.
- The charge ensemble is thus a special kind of dipolarity. It exhibits asymmetry of opposite charges in its virtual photon exchange *a priori*.
- Both the bare charge in the middle and the surrounding polarized charge are *infinite*. Their difference is *finite*. The two infinite charges will forever sustain the finite difference, so long as the polarization ensemble is not forcibly altered.

INFINITE VIRTUAL CHARGES
(-) + (-) (-) (-) (-)
OBSERVE DIFFERENCE THROUGH SHIELDING
INFINITE BARE CHARGE
"ISOLATED" CHARGE

* For a beautiful discussion of the polarization and much more, see Steven Weinberg, Dreams of a Final Theory, Vintage Books, Random House, 1993.

A charge is a dipolar ensemble (2)

- Because of the broken symmetry of the infinite dipolarity, a set of energy flows with finite intensity must therefore emerge from the polarized charge ensemble. Playing with charges means playing with finite differences between appositive infinite energy flows.
- Because the fundamental charges have infinite charge and infinite energy, the flow of energy from the charge will be sustained forever.
- The broken symmetry of opposite charges was proven by Wu et al. in 1959, and a Nobel Prize was awarded to Lee and Yang the same year.

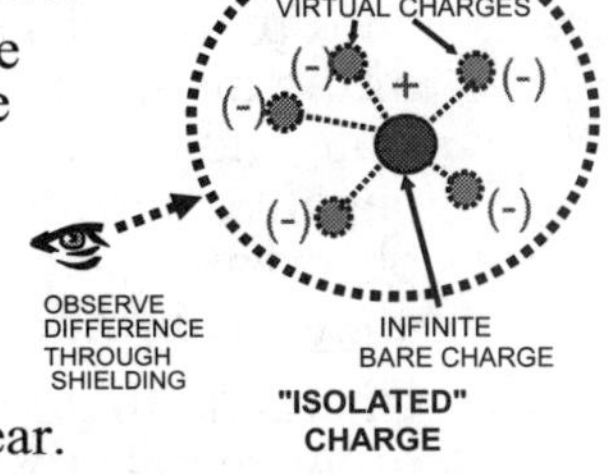

- The *observable* flow of EM energy is from *virtual* energy that has "become observable" (been reordered, coherently integrated to the quantum level, and re-emitted as quanta).

Solution to the source charge problem

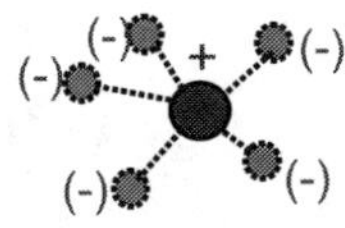

- Opposite charges also are a broken symmetry in the fierce vacuum's virtual particle exchange.
- Continually absorbs *virtual* EM energy, transduces it into *observable* EM energy (photons), and then emits the observable photons in all directions in 3-space.
- The source charge and any dipolarity continuously consume positive entropy of the virtual state and convert it to negative entropy of the observable state.
- The charge falsifies the present second law of thermodynamics, requiring its revision, and is an example of a real system continuously producing negative entropy.*

---

* First such known physical system, although permissibility was theoretically demonstrated by D. J. Evans and Lamberto Rondoni, "Comments on the Entropy of Nonequilibrium Steady States," J. Stat. Phys., 109(3-4), Nov. 2002, p. 895-920.

Where EM energy comes from:

- Every joule of EM energy comes from the vacuum, via broken symmetry of source charges (dipolarities, considering polarized vacuum).
- Charge consumes positive entropy (breaks symmetry) of the disordered virtual state and converts it to negative entropy of the observable state. There is a *hierarchy* of symmetries.*

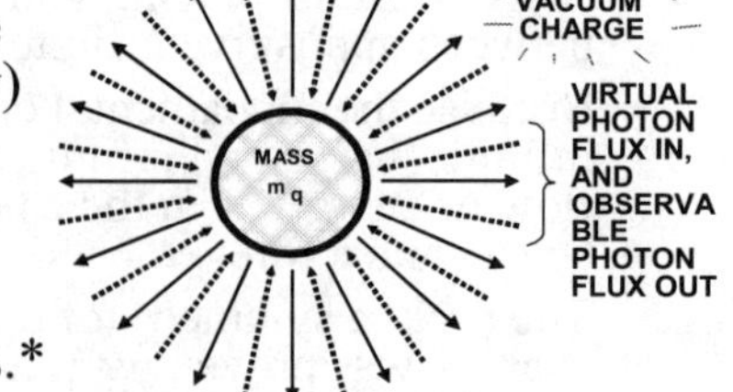

- Charge is first known physical system to continuously produce negative entropy, as theoretically possible**.

---

* This violates Klein geometry; Leyton geometry must be used. See Michael Leyton, A Generative Theory of Shape, Springer-Verlag, Berlin, 2001.

** Evans and Rondoni, "Equilibrium microstates which generate second law violating steady states," Phys. Rev. E, Vol. 50, 1994, p. 1645-1648.

The magic reordering (negative entropy) process

$$m^*(t) = m(t_0) + dm_1 + \quad dm_2 + dm_3 + \ldots$$

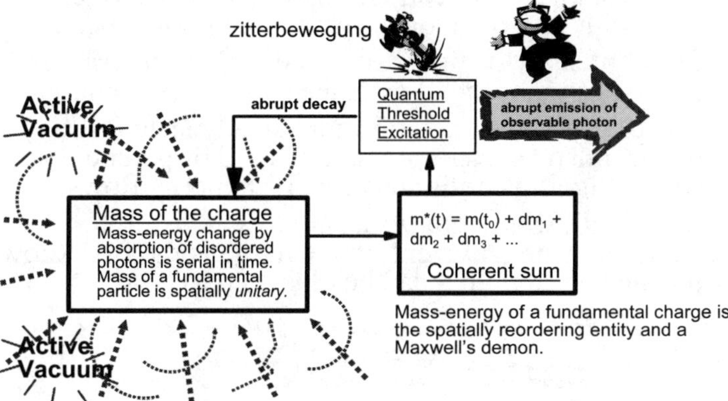

Mass-energy of a fundamental charge is the spatially reordering entity and a Maxwell's demon.

The fundamental charged particle continually absorbs disordered virtual state EM energy and re-emits it as ordered observable EM energy

- Each increment absorbed of virtual energy produces a virtual-change of the charge's <u>*already unitary*</u> mass.*
- The serial differentials of its mass-energy are *spatially re-ordered energy*. Each is a change to what is there before.
- The differentials thus *coherently sum*, so a single, *unitary* differential of virtual mass-energy (its excitation) grows.
- When the quantum threshold is reached, there is enough new mass-energy to produce a real, observable photon .
- Zitterbewegung** shakes out an observable photon, emitted from the particle in its abrupt excitation decay.
- <u>*Feynman ratcheting:*</u> The virtual excitation process continues again, and the process reiterates unceasingly.

---

*This is a true Maxwell's Demon, totally destroying the present second law of thermodynamics.

** Zitterbewegung is a local circulatory motion of the electron presumed to be the basis of the electron spin and magnetic moment. One physical interpretation is given by David Hestenes, "The Zitterbewegung Interpretation of Quantum Mechanics," <u>Found. Phys</u>., 20 (10), 1990, p. 1213-1232.

Virtual photon interactions reveal secret of negative entropy

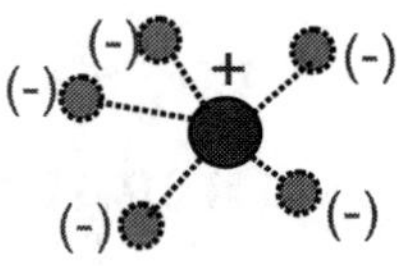

- Observable photons from source charge continuously radiate away at speed of light.
- They establish and continuously replenish associated EM fields and potentials, spreading at light speed.
- Point intensity of fields and potentials is *deterministic*, as a function of radial distance from source charge.
- The charge is a nonequilibrium steady state (NESS) deterministic system. A "static" field or potential is a set of ongoing continuous free EM energy flows.
- Such a NESS system *theoretically* can exhibit continuous negative entropy. Charge is the first known physical system that can be observed to do so.**

---

* See Evans and Rondoni, "Equilibrium microstates which generate second law violating steady states," Phys. Rev. E, Vol. 50, 1994, p. 1645-1648.

** See T. E. Bearden, Fact Sheet 2003-01, "The Source Charge Problem: Its Solution and Implications," www.cheniere.org, Aug. 18, 2003.

Charge falsifies the old second law of thermodynamics to any macroscopic level since the fields and potentials are deterministic (1)

- Charge exhibits continuous negative entropy production.
  - Falsifies second law, which is based on Klein's 1872 geometry* and group theoretic methods.
  - Validates theoretical proof that continuous production of negative entropy is possible, as shown by Evans and Rondoni.**

LEYTON

- In Klein geometry, a broken symmetry at one level loses the information at that level, reducing the overall symmetry to that of the lower level. *This is a positive entropy process because the ordering is decreased.*
- So $dS \geq 0$ is arbitrarily "wired in" to Klein geometry!

---

*Felix Klein, "Vergleichende Betrachtungen über neuere geometrische Forschungen," 1872.

**D. J. Evans and Lamberto Rondoni, "Comments on the Entropy of Nonequilibrium Steady States," J. Stat. Phys., 109(3-4), Nov. 2002, p. 895-920.

Charge falsifies the old second law of thermodynamics to any macroscopic level since the fields and potentials are deterministic (2)

LEYTON

- In Leyton's object-oriented geometry,* a broken symmetry at one level generates a new symmetry at the next higher level, with a layer incorporating all information of the lower levels. *This is a negative entropy process because the ordering is increased.*
- Leyton's extended geometry and deeper group theoretical methods inherently falsify the present second law of thermodynamics by including negative entropy and the negative entropy action of the source charge.
- Leyton's work is a new revolution in physics, as fundamental as was the original prediction of asymmetry by Lee and Yang, awarded the Nobel Prize in 1957.

* Michael Leyton, A Generative Theory of Shape, Springer-Verlag, Berlin, 2001

Breaking the zero-symmetry state of the disordered vacuum (1)

- Vacuum fluctuations are disordered and subquantal, overall (statistically).
- Hence the overall ordering (symmetry) of the vacuum may be taken as zero. This zero symmetry is a *specific symmetry state of the overall vacuum*.
- But in these fluctuations, local polarizations (opposite charges) continually appear and disappear.
- While a local polarization exists, it is a "broken symmetry" of the vacuum state. Hence it is a local violation of the "zero symmetry" of the overall vacuum, and the usual transient fluctuation theorems apply.*

* E.g., see D. J. Evans and D. J. Searles, "Equilibrium microstates which generate second law violating steady states," Phys. Rev. E, Vol. 50, 1994, p. 1645-1648.

Breaking the zero-symmetry state of the disordered vacuum (2)

- Equilibrium is the state of maximum entropy of a system.
- In a system in statistical equilibrium, local transient statistical fluctuations continually occur.
- In a local transient fluctuation
  - Equilibrium is broken and entropy is momentarily reduced.
  - The Second Law of thermodynamics is violated and such fluctuation obeys a transient fluctuation theorem.*
  - Violation can last appreciably — e.g., up to two seconds.**
- Local nonequilibrium steady state interactions with the observable state (as with a fundamental charged particle) can violate the second law of thermodynamics.***

---

* E.g., see D. J. Evans and D. J. Searles, Phys. Rev. E, Vol. 50, 1994, p. 1645-1648.
** G. M. Wang et al., Phys. Rev. Lett., 89(5), 29 July 2002, 050601.
*** D. J. Evans and Lamberto Rondoni, J. Stat. Phys., 109(3-4), Nov. 2002, p. 895-920.

Every charge consumes positive entropy of *virtual* state and produces negative entropy of *observable* state

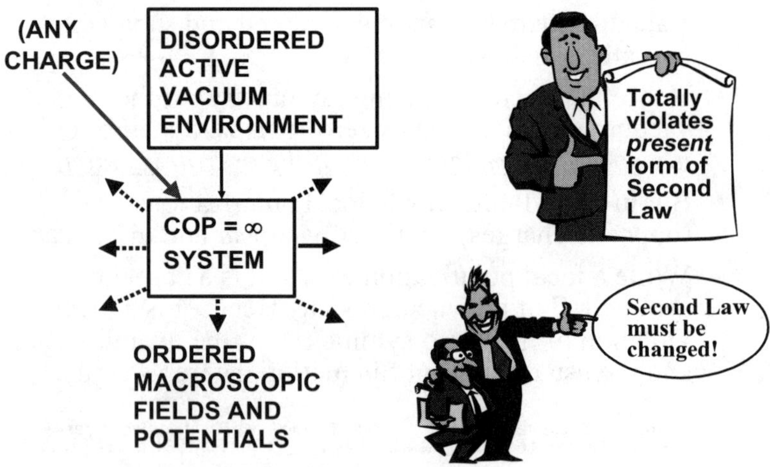

Biochemistry and EM models (1)

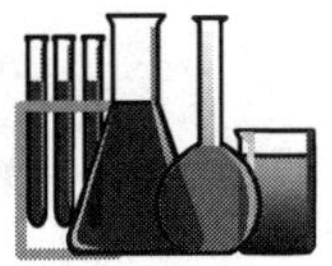

- Medical scientists analyze such things as teratomas in terms of biochemistry and the biochemical model.
- An assembly of charges and charge distributions, with their associated fields and potentials, drives chemistry at base level.

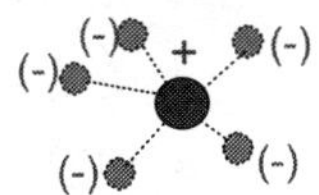

- The force dynamics of the charge assembly are generated by a corresponding assembly of spacetime curvatures and their dynamics (a spacetime curvature engine).
- This *precursor engine* is the force-free cause that is driving the charge assembly and generating its dynamics, producing all forces in the matter and therefore the chemistry ongoing in that matter.

Biochemistry and EM models (2)

- Force dynamics of the charge assembly are generated by a precursor force-free spacetime curvature engine.
- This precursor engine is *the force-free cause that is driving the charge assembly and generating its dynamics, producing all forces in the matter and therefore the chemistry ongoing in that matter*.
- Precursor fields and potentials are not modeled in classical U(1) EM theory or quantum electrodynamics.
- CEM/EE model is 140 years old and seriously flawed.
- Precursor force-free EM fields and potentials — and their interaction with charges to produce the force fields and dynamics — can be modeled in unified field theory (such as that of Sachs and that of Evans in O(3) electrodynamics).

Biochemistry and EM models (3)

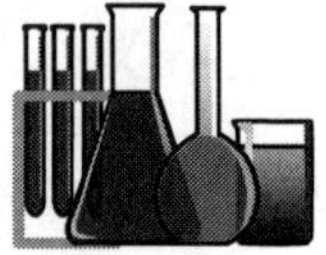

- Conventionally a causative EM field in space is "defined" as the *diversion from* a spatial *force* field by and during its interaction with charge.
- But force exists only in and of mass. Mass is a component of force, by $F \equiv dp/dt = d/dt(mv)$.
- *There is no force field in space*. Assuming one in space substitutes "effect in matter" for "matter-free ST curvature cause" — a logical non sequitur.
- CEM/EE also assumes charged matter (the material ether) filling all space, and calculates and uses only the force field existing in that assumed charged matter.
- CEM/EE does not and cannot calculate the *force-free precursor EM field* in space — the actual EM field.

How CEM/EE erroneously assumes force fields in space

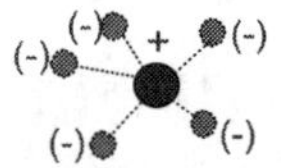

- Jackson* shows how electrodynamicists simply avoid (and dispose of) the problem of a force-free, massless EM field in space, rather than a force field:

> *"Most classical electrodynamicists continue to adhere to the notion that the EM force field exists as such in the vacuum, but do admit that physically measurable quantities such as force somehow involve the product of charge and field."*

---

* J. D. Jackson, Classical Electrodynamics, 2nd Edn., John Wiley & Sons, New York, 1975, p. 249. Jackson is one of the leading classical electrodynamicists and either highlights or clarifies many issues.

U.S. scientists do not calculate the EM field in space, but only in charged matter

- So CEM/EE still erroneously assumes a material ether and force fields in space. Ignores the precursor field.
- *There is not now and there never has been an EE department, EE professor, or EE textbook that teaches how to calculate the (precursor) EM field in space.*
- This is a known major flaw in the CEM/EE model. It has been pointed out in one fashion or another by many physicists such as Wheeler and Nobelist Feynman.
- The CEM/EE model is seriously flawed, and the flaws emerge in biochemistry and medical science.
- At least Jackson clearly states the gross error that is still accepted and used by CEM/EE.

* J. D. Jackson, Classical Electrodynamics, 2nd Edn., John Wiley & Sons, New York, 1975, p. 249.

Correction to basic mechanics and a solution to Feynman's problem of defining force*

- $F \equiv \partial/\partial t(mv)$, so *mass is a component of force*.
- Force and force fields exist only in matter, not in massfree space.
- There is no separate "massless vector force in space" acting upon a separate mass.
- Thus in EM models, cannot legitimately use force fields as a *causative agent in space*, as present models do.
  - EM field in space is a *massless, force-free* causative agent (curvature of spacetime) — a *precursor* to force field in matter.
  - Force-free EM field's ongoing interaction with charged matter *identically is a force field in that matter*. Force is the interaction effect in and of the charged matter itself, not the cause.

* Richard P. Feynman, Robert B. Leighton, and Matthew Sands, The Feynman Lectures on Physics, Addison-Wesley, Reading, MA, Vol. 1, 1964, p. 2-64, 12-2; Vol. 2, p. 1-3.

What physics has to say about EM field in space

Richard Feynman

- Feynman* clearly states that the EM field in space is not a force field.
- It is a characteristic of space that *produces* a force field when and if it interacts with charged matter.

> "*...the existence of the positive charge, in some sense, distorts, or creates a "condition" in space, so that when we put the negative charge in, it feels a force. This potentiality for producing a force is called an electric field.*"

> "*We may think of E(x, y, z, t) and B(x, y, z, t) as giving the forces that would be experienced at the time t by a charge located at (x, y, z), with the condition that placing the charge there did not disturb the positions or motion of all the other charges responsible for the fields.*"

---

*Feynman, Leighton, and Sands, The Feynman Lectures on Physics, Addison-Wesley, Reading, MA, Vol. 1, 1964, p. 2-4; Vol. 2, p. 1-3.

Maxwell's theory was curtailed*

"*[T]he A field [for the potentials] was banished from playing the central role in Maxwell's theory and relegated to being a mathematical (but not physical) auxiliary. This banishment took place during the interpretation of Maxwell's theory... by Heaviside... and Hertz. The 'Maxwell theory' and 'Maxwell's equations' we know today are really the interpretation of Heaviside... Heaviside took the 20 equations of Maxwell and reduced them to the four now known as 'Maxwell's equations'.*"

COMMENT: Maxwell's primary theory was 20 quaternion-like equations in 20 unknowns.**

---

*Terence W. Barrett, "Electromagnetic Phenomena Not Explained by Maxwell's Equations," A. Lakhtakia, ed., Essays on the Formal Aspects of Electromagnetics Theory, World Scientific Publishing, River Edge, NJ, 1993, p. 11 and generally p. 6-86.

** James Clerk Maxwell, "A Dynamical Theory of the Electromagnetic Field," Royal Society Transactions, Vol. CLV, 1865, p 459

Necessary EM corrections*

- Higher symmetry, non-Abelian electrodynamics models have been developed in particle physics.
- The new approach extends these models to also incorporate general relativity, thus producing a unified field theory*.
- In the new approach, spacetime curvature engines are perfectly natural and understandable, and can be directly modeled and engineered by advanced theorists.
- The theory is sufficiently advanced and robust that technology and engineering can now advance.
- This would be a great leap forward for biochemistry and medical science, particularly in therapeutic methods.

* Sachs's unified field theory implemented by O(3) electrodynamics developed by Evans presently is suitable to use. Numerical methods are often required, particularly when asymmetry is invoked, Lorenz symmetry is violated, and Lorentz-invariant equations cannot describe all the dynamics and results.

A fundamental charged particle is a Feynman ratcheting system* (1)

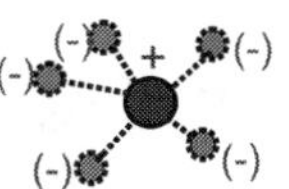

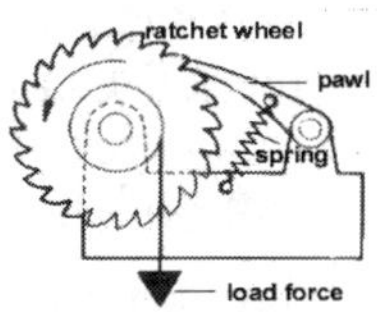

- Assume fundamental charge at ZPE equilibrium state (lowest quantal level).
- Disordered virtual photons from vacuum are absorbed serially by the fundamental charged particle. This is a reordering of the absorbed energy in 3-space and time.
  - Mass of a fundamental particle is already unitary in 3-space.
  - The unitary mass increases by little serial differential increases.
  - The serial integration completes the negentropy process.
- On the mass-energy, the same type of virtual energy differential thus appears in serially increasing order. The mass-energy of the charge at lowest quantum level (ZPE equilibrium state) has a series of *coherently integrating* differential increases in the virtual state.

* E.g., see Jack Denur, "Modified Feynman ratchet with velocity-dependent fluctuations," posted on web at http://www.ipmt-hpm.ac.ru/SecondLaw/down/170.pdf .

A fundamental charged particle is a Feynman ratcheting system (2)

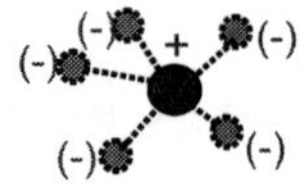

- When the integrating differential *virtual state* mass-energy reaches the next higher quantum energy level, the violent zitterbewegung (jitter) of the absorbing charged particle abruptly initiates emission of an observable photon.
  - The emission abruptly decays the integrated excited state of the charged mass back to zero reference (i.e., the ZPE equilibrium state at base level).
  - Process continually reiterates since virtual disordered photons are continually absorbed in serial order by each charge, driving the process. This is a *Feynman ratcheting* effect.

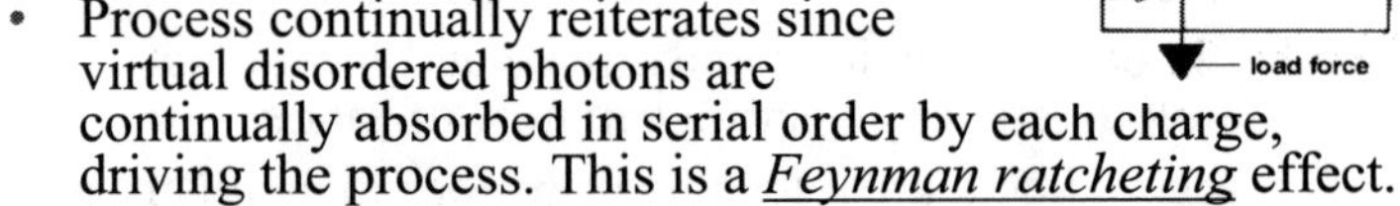

- The charge continually emits real observable photons in all directions, without any *observable* energy input. The input is a "sorted and reordered" *virtual state* energy input.The charge is a Feynman ratchet system, producing continuous negative entropy from positive energy and its "ratcheting" action.

Consistent with ZPE concepts, thermo-dynamics, and EFTV concepts (1)

- A charge at zero absolute temperature has QM movements and energy. This is the lowest quantal level, called "zero-point energy" state of the observed charged particle.
- The interaction of the charge — its absorption of disordered *subquantal virtual energy fluctuations*, their conversion to differentials of unitary mass-energy with serial ordering, and coherent integration of the mass-energy differentials — excites the charge to the *next higher* quantum level, moving it out of ZPE-level equilibrium.
- The excited (potentialized) charge is *then* permitted to decay back to ZPE level, emitting an observable photon.

Consistent with ZPE concepts, thermodynamics, and EFTV concepts (2)

- Continual iteration of this "ratcheting" action — converting absorbed virtual state energy to emitted observable state energy and reiterating — provides continual emission of real photons by the charge, where energy of the emitted photons has been "lifted" from the virtual state vacuum.

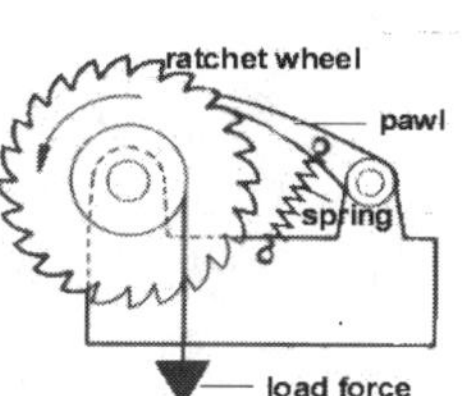

- There is no "observable" (quantal) energy input because the required energy input is in *subquantal* form.
- ZPE consistency is upheld, thermodynamicists are happy, and EFTV enthusiasts are happy.
- Electrodynamicists should be happy that the long-vexing source charge problem is *finally* resolved!

Nonequilibrium steady state system

- A nonequilibrium steady state (NESS) system can:
  - Self-order.
  - Self-oscillate or self-rotate.
  - Output more energy (as powered loads, etc.) than what the operator alone inputs. The excess energy input comes directly from the active environment, as free *energy transfer*.
  - Power itself and its load (its associated observable fields and potentials in space). The input energy comes from the active vacuum environment, via the source charges, and the operator need not pay for it.
  - Exhibit negative entropy.
- The charge with its subquantal vacuum energy input and its observable EM energy output is a NESS system providing freely flowing EM energy from the vacuum.

PRIGOGINE

All EM energy is free: EM fields and potentials are sets of ongoing and free EM energy flows (1)

- The charge is a nonequilibrium steady state (NESS) system. So are its "static" fields and potentials.
- Charge continually *consumes* vacuum's positive entropy at the virtual state level, and *produces* negative entropy output at the observable level.
  - Transduces absorbed disordered virtual state energy (from vacuum subquantal fluctuations) into observable state energy (real observable photons, real EM fields).
  - Falsifies the old second law of thermodynamics, requiring its correction and extension.
- *Continuously* produces negative entropy observably, as shown theoretically possible by Evans and Rondoni.*

---

* D. J. Evans and Lamberto Rondoni, J. Stat. Phys., 109(3-4), Nov. 2002, p. 895-920.

All EM energy is free: EM fields and potentials are sets of ongoing and free EM energy flows (2)

- A truly "static" EM field does not exist**;
- Every EM field is a set of NESS flows of energy*** from its associated source charge or dipole.
- All we have to do is learn how to properly intercept, collect, and freely utilize the freely flowing EM energy.
- The collection and utilization process is an asymmetrical process *a priori*.
- Lorentz made the CEM/EE equations symmetrical so that simpler Lorentz-invariant equations could be used.
- The first requirement for COP>1.0 EFTV systems is to violate Lorentz symmetry in one's circuits, so that non-invariant equations cannot model their operation

---

* See Tom Van Flandern, Phys. Lett. A, 250, Dec. 21, 1998, p. 1-11.

** E.T. Whittaker, 1903 and 1904 (ibid.).

*** Many methods found by inventors are given in T. E. Bearden, Energy from the Vacuum: Concepts and Principles, Cheniere Press, 2002. For rigorous proof that violating the Lorentz symmetry condition allows EFTV systems, see M.W. Evans et al., Physica Scripta, Vol. 61, 2000, p. 513-517.

Importance of solution to source charge problem (1)

- The source charge *transduces* disordered virtual fluctuation energy from the vacuum into real, usable EM energy flows comprising its fields and potentials.
- Extracting usable energy from the vacuum, any amount, anywhere, is straightforward. Every charge already does it freely.*

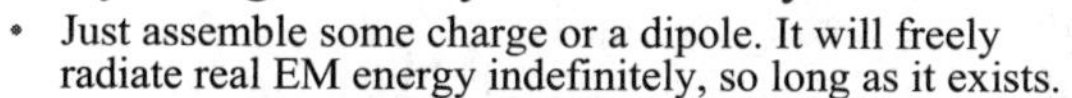

  - Just assemble some charge or a dipole. It will freely radiate real EM energy indefinitely, so long as it exists.
  - Use a collecting circuit having overall broken Lorentz symmetry.
  - Potentialize the circuit with charges pinned so dq/dt = 0.
  - When current flows and the collected energy is dissipated in the load, the source of potential and energy flow must not be connected into that external circuit as a load to be destroyed.

(-) (-) + (-) (-) (-)

---

* Again, for examples of such systems produced by inventors, see T. E. Bearden, Energy from the Vacuum: Concepts and Principles, Cheniere Press, 2002. E.g., any EE department can readily build the Takahashi magnetic Wankel engine (suppressed by the Japanese Yakuza), which kills its own back mmf and thus produces a nonconservative rotation force, continually driving the engine and powering its load. It may also be readily made self-powering.

Importance of solution to source charge problem (2)

- There is no scientific excuse whatsoever for an electrical engineering that implicitly assumes that
  - Every EM field and potential (and every joule of EM energy) is freely created by the source charge *from nothing at all*.
  - EM fields and potentials are "static".
  - Circuits must be built Lorentz-symmetric so as to deliberately self-destroy their own dipolar source of free EM energy from the vacuum.
  - EM energy occurs in "chunks" called joules. Instead, it occurs in free flows of EM energy — in *joules per second*.
  - EM fields in space are force fields.
- We must rethink our concept of EM energy and its use. The old "first principles" are seriously flawed.

(-) (-) + (-) (-) (-)

EM energy is free, but we've not yet learned to use it properly! (1)

- Every charge is already a "free gusher of EM energy" that will flow forever if the charge is left undisturbed.
- Every "static" EM field or potential is thus a set of ongoing steady-state EM energy flows.*
  - Van Flandern's analogy applies.**
  - A "static" EM field or potential is like an *unfrozen* waterfall, rather than like a *frozen* waterfall as erroneously portrayed.
- Having the source connected to its external circuit while current is flowing destroys the source dipolarity — and the free flow of energy from the vacuum — faster than the load is powered.

---

* E. T. Whittaker, Math. Ann. Vol. 57, 1903, p. 333-355 and Proc. Lond. Math. Soc., Series 2, Vol. 1, 1904, p. 367-372 proved that any EM field or potential is a set of ongoing bidirectional EM energy flows.

** Tom van Flandern, Phys. Lett. A 250, Dec. 21, 1998, p. 8-9.

EM energy is free, but we've not yet learned to use it properly! (2)

- Imposition of Lorentz symmetry to allow simpler (Lorentz invariant) equations is wrong. Building only symmetric circuits limits them to COP < 1.0, and keeps the meter on the gas pump and the meter on one's house.
- Having the dipolar source connected to its external circuit while current is flowing:
  - Destroys source dipolarity — and the free flow of energy from the vacuum.
  - Uses half the free EM energy collected on circuit charges to destroy source dipole and cut off free energy flow from vacuum.
- We have not learned how to intercept work-free potential energy flow, *collect it for free without current flowing*, then disconnect source and allow current to flow, freely dissipating the freely collected energy to power external loads "for free".

Energy, Spacetime, Vacuum VPF

- There really has been no satisfactory definition of energy.
- *Energy change* is measured, not *energy per se*.
- Energy can be defined in either of two ways:
  - Spacetime is identically energy.
  - Virtual particle flux (of the vacuum) is identically energy.
  - In either case, the energy density is enormous.
- A "change of energy" thus means two things:
  - A change (i.e., distortion or curvature) of spacetime.
  - A change in the virtual particle flux (VPF) of the vacuum.
- The "field" is a region of distorted spacetime and altered VPF.
- Fields and potentials decompose into *sets of energy flows*. So they are regions of flowing changes in spacetime distortion, and flowing changes in vacuum VPF.

The entropy concept implicitly assumes its own contradiction (1)

- Price points out the problem*:

  "A century ago Ludwig Boltzmann and others attempted to explain the temporal asymmetry of the second law of thermodynamics. The hard-won lesson of that endeavor — a lesson still commonly misunderstood — was that the real puzzle of thermodynamics is not why entropy always increases with time, but why it was ever so low in the first place."

- For a system initially in equilibrium, the concept of further entropy production implicitly assumes a previous negentropic process has occurred to provide additional usable (controlled) order (usable energy) in the system, that is to be disordered and decontrolled.

* H. Price, Time's Arrow Today, S. F. Savitt, Ed., Cambridge University Press, 1995, p. 66-94.

The entropy concept implicitly assumes its own contradiction (2)

- Scientists balk at the necessity for initial negative entropy production, and largely avoid facing it.
- Usually they arbitrarily invoke positive entropy occurring in the external environmental system, and thus furnishing the energy — to the system — in work-free manner but with work in the environment.
- This just moves the problem to an outside system (the environment) and does not solve it.
- Scientists stared at the necessity for equal negative entropy first occurring, and then walked away, rather than dealing with it*.

* By his hierarchies of symmetry and change to geometry, Leyton solved the problem. See Michael Leyton, A Generative Theory of Shape, Springer-Verlag, Berlin, 2001.

Importance of Leyton geometry*

- We also must upgrade thermodynamics by adopting Leyton geometry instead of the limited 1872 Klein geometry.
- Leyton's work proves negative entropy as the other "missing" half of the Second Law of thermodynamics — the half which is transduction of entropy into negative entropy, as exhibited by the source charge and its deterministic EM fields and potentials and their energy.

Dr. Michael Leyton

* See Michael Leyton, A Generative Theory of Shape, Springer-Verlag, Berlin, 2001. Leyton's object-oriented geometry uses more advanced groups and group operations than the old 1872 Klein geometry. Klein geometry has entropy "welded in", so to speak. In Leyton geometry, this is not the case; rather, negative entropy is also included, particularly in Leyton's hierarchies of symmetry.

The Leyton effect:

- In Klein geometry, broken symmetry at a given level loses the information at that level and *reduces* the overall group symmetry. This is basically a "wired-in entropy operation."
- In Leyton geometry, broken symmetry at a given level automatically creates a new symmetry at the next higher level, with a layer that captures all the lower level information. Hence a broken symmetry *increases* the overall group symmetry. This is a negative entropy operation *a priori*.
- Leyton geometry thus produces a *hierarchy of symmetries*, missing from Klein geometry.

Leyton's geometry solved the overall problem of thermodynamics

- The overriding thermodynamics problem is: *If the second law is correct, and entropy can only stay the same or increase, then how did the entropy of the universe ever get so low in the first place? And how is it so low now?*
- How does a system, once reaching equilibrium (maximum entropy), ever depart from equilibrium to lower entropy?
- The old second law also does not consider negative energy and negative energy EM fields — the so-called "dark energy" of the universe. It does not account for negative mass matter — the so-called "dark matter" of the universe.
- Problem recognized — and unsolved — for a century.
- It cannot be solved unless there is also a law of negative entropy production, which immediately requires hierarchies of symmetry. Leyton's work provides that missing law.

Correction of the second law of thermodynamics (1)

- We have proposed the following correction:
  "For a system initially in equilibrium, first some negative entropy is produced in the system, departing it from equilibrium and lowering its entropy so it has some controlled order (energy) to be decontrolled and disordered. In interactions thereafter, the entropy remains the same or increases, unless additional interactions producing negative entropy arise and interfere."
- This statement is consistent with experiment (e.g., Wang et al.*), Leyton geometry,** and the source charge's ratcheting production of its associated EM fields and potentials and their energy.***

---

* G. M. Wang, E. M. Sevick, Emil Mittag, Debra J. Searles, and Denis J. Evans, "Experimental Demonstration of Violations of the Second Law of Thermodynamics for Small Systems and Short Time Scales," Phys. Rev. Lett. 89(5), 29 July 2002, 050601.

** Michael Leyton, A Generative Theory of Shape, Springer-Verlag, Berlin, 2001.

*** T. E. Bearden, Energy from the Vacuum: Concepts and Principles, 2002.

Correction of the second law of thermodynamics (2)

- Present second law is ($0 \leq dS < +\infty$), which states that the system entropy can only stay the same or increase.
- System in equilibrium is in state of maximum entropy. Thus to produce entropy, the system must *first* depart from equilibrium, lowering its entropy.*
- This is a negative entropy operation ($-\infty < dS \leq 0$) and it produces a "potentialized" system or "excited" system not in equilibrium.*
- The excited or potentialized system can then *entropically* decay back to equilibrium condition, if no further excitation or potentialization (negative entropy operation) occurs to interrupt it. This is ($0 \leq dS < +\infty$) — the present old second "half-law".

•If the energy is furnished to the system in the same form as the system's potential energy that is changed in magnitude, no work is done. This is gauge freedom, which is inherently negentropic. If the energy is furnished in a different form, work is required in order to change the form of the input energy.

Correction of the second law of thermodynamics (3)

UNACCOUNTED NEGATIVE ENTROPY OPERATION

- The two serial actions of negative entropy and positive entropy form a *ratcheting* process, similar to Feynman's ratchet.
- The second law has always been an oxymoron implicitly assuming its own contradiction has first occurred but has not been accounted.
- The new, complete second law is just ($-\infty < dS < +\infty$), and it states that in an interaction the system may produce negative entropy, positive entropy, or stay at the same entropy level — all depending upon the nature of the interaction.

Correction of the second law of thermodynamics (4)

W = Vq

- New, corrected second law is ($-\infty < dS < +\infty$).
- In macroscopic systems, often potentialization and dissipation are serial, discrete operations. In that case:
  - First ($-\infty < dS \leq 0$) occurs separately.
  - Then ($0 \leq dS < +\infty$) occurs separately.
- So in this special case the new second law is: $(-\infty < dS < +\infty) = (-\infty < dS \leq 0) + (0 \leq dS < +\infty)$
- The implicitly assumed unaccounted negative entropy operation is clear, as is the proper meaning of the old second "half-law". It was just entropic decay back to equilibrium, of a *previously excited system*.
- Now electrical engineering can proceed with *negative entropy engineering*, directly.

Correction of the second law of thermodynamics (5)

W = Vq

- The new second law and revised first law allow self-powering electrical power systems taking their input energy freely from the vacuum. The source charge is one such self-powering EM system.
- From any fixed source of potential V, given sufficient q any amount of potential energy W can be freely collected, by W = Vq, or for a given q by using repetitive collection-dissipation with the source connected *only in collection*.
- In short, just use a similar "ratcheting" operation as shown by the source charge.

Implications for electrical power systems: A permanent solution to the energy crisis (1)

- For $(-\infty < dS < +\infty) = (-\infty < dS \leq 0) + (0 \leq dS < +\infty)$:
  - No law of nature states we ourselves must pay for the extra potentialization energy $(-\infty < dS \leq 0)$. Pure asymmetric regauging (voltage amplification) can supply it.
  - Conservation of energy merely requires that *something* (such as the environment) must furnish it, since we dissipate it by $(0 \leq dS < +\infty)$. Any source of potential can give it freely.
  - We *can* insist on furnishing it by brute force, paying for it.
  - Or we *can* allow nature to furnish it freely — *including from the vacuum*.
- EM energy in nature occurs in steady, inexhaustible free energy flows from all source charges and dipoles. *This allows use of work-free asymmetric regauging*.

Implications for electrical power systems: A permanent solution to the energy crisis (2)

- We may therefore *freely* collect extra EM potential energy from the gushing energy flows coming from the circuit charges or system charges.
- By adroit switching, this potential energy can be dissipated in an external load to "power" it, *without* using half of it to "kill the source dipolarities".
- We need pay only a little for the "switching and control" (S&C) energy dissipated.
- From the output load energy, we can extract that small S&C energy via positive feedback from the load output to the S&C unit, so the S&C unit is a small part of the "powered load".
- In that way, the working system is also self-powered by energy freely extracted from the vacuum.

Example: Any EE department can build a self-powering magnetic Wankel engine (1)

- It is a rail gun wrapped nearly in a complete circle, reducing back mmf to a very small open section.
- A pole piece with coil covers gap. A very small trickle current is in the coil.
- Current is sharply broken as rotor enters gap.
- Lenz law surge of current creates magnetic field that momentarily overrides back mmf field.
- Rotor passes through a zero "back mmf" field; even a forward mmf field.
- Pay a little, get engine with no back mmf, violating Lorentz symmetry. COP > 1.0 easily.

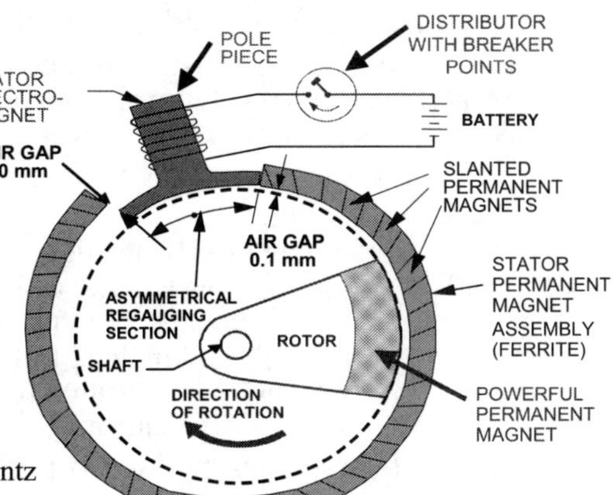

Example: Any EE department can build a self-powering magnetic Wankel engine (2)

- Easily made self-powering by adding a small DC generator as part of shaft load.
- Once at speed, an automatic control circuit switches battery out of loop and switches in the generator.

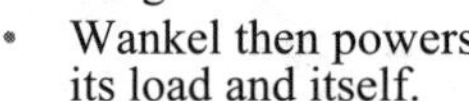

- Wankel then powers its load and itself.
- It is a nonequilibrium steady state (NESS) system, permitted to be self-powering.
- All energy input is furnished by the vacuum.

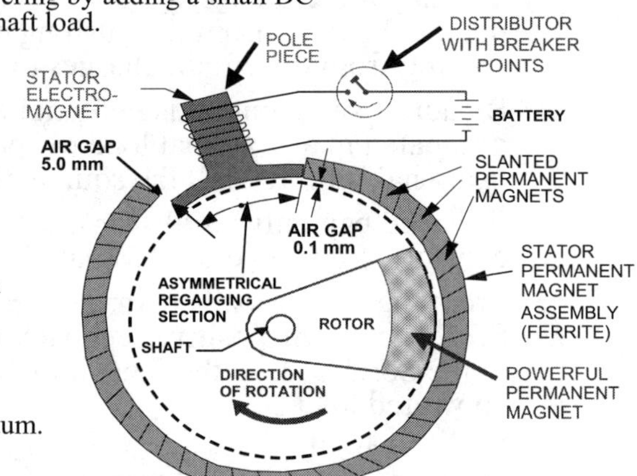

Since EM fields and potentials are sets of energy flows:

- All operating EM systems are systems far from equilibrium. Consider steady state condition:
- These EM systems are nonequilibrium steady state (NESS) thermodynamic systems, freely receiving their processed energy from the vacuum environment.
- Such a system has five "magic" characteristics. It can:
    - Self-order.
    - Self-rotate or self-oscillate.
    - Output more energy (as work) than the operator inputs (the excess energy input comes from the environment).
    - Power itself and its load (all the energy input comes from the active environment).
    - Exhibit negative entropy.
- It is time we learned to properly *utilize* our NESS EM electrical power systems, rather than *killing* them.

Why Yakuza suppresses the Wankel magnetic engine and several others

- About two years from now, with scalar interferometry the Yakuza plans to disable the centralized electrical power system of the U.S., to never work again.
- As total economic collapse looms, U.S. will finally boot its scientific community hard, and there will be desperate search for a quick solution.
- Through the Japanese government the Yakuza will offer Japanese self-powering systems as the quick answer.
  - Draws off all available desperation funding that is "last ditch" U.S. effort to prevent total economic collapse.
  - By delivering only a few systems and very slowly, that "last ditch" U.S. science effort will be thwarted.
- As a result, there will be a total U.S. economic collapse, which — coupled with conversions, nuclear explosions in cities, anthrax attacks, etc. — will finish off the U.S.
- Yakuza then kills remaining members of U.S. population.

Correcting an erroneous interpretation of the *first* law of thermodynamics (1)

- The First Law is *erroneously* thought to define any change in magnitude of an external parameter of the system as work *a priori.**
- Potential energy of a system in a given form is such a parameter.
- Work rigorously is *the change of form of energy*, but not change of *magnitude* of energy remaining in the same form. The latter requires no work.
- Mere change of magnitude of potential energy of a system is *regauging*, and it is work-free by the gauge freedom axiom of quantum field theory.

* E.g., see Ralph Baierlein, Thermal Physics, Cambridge University Press, 1999, p. 2.

Correcting an erroneous interpretation of the *first* law of thermodynamics (2)

- Present *interpretation* of the first law excludes gauge freedom — a major non sequitur requiring correction.
- If energy is input — by the external environment or fixed source of potential — to the system in the *same form as* the external parameter's energy, then no work is required.
- In that case, the change of the external parameter's magnitude is by work-free asymmetric regauging.
- If energy is freely input to the system in a *different* form than the external parameter's energy, then work must be done to change the form of the input energy, before it can change the magnitude of the external parameter.
- Properly switched and controlled, a potentializing source of freely flowing EM energy can furnish an unlimited amount of freely collected potential energy, *indefinitely*.

Correcting an erroneous interpretation of the *first* law of thermodynamics (3)

- $W = \phi q$ shows that any amount of energy W can be freely collected from a fixed scalar potential ϕ, by charges q, if the amount of charge q intercepting the energy (deviating it from the flows comprising ϕ) is freely variable.
- If the amount of charge q is also fixed, any amount of energy W can still be freely collected from ϕ by ratcheting iterative static collection of energy *into* the circuit and dynamic dissipation of energy *from* the circuit to power loads.

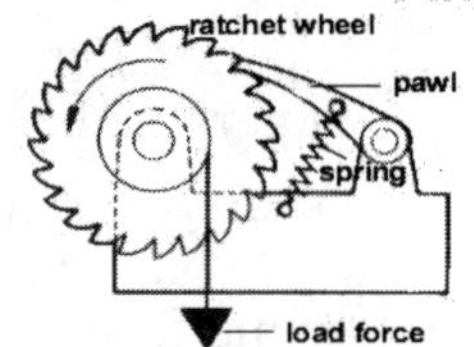

- Similarly, any amount of emf force can be freely collected from a fixed E-field by $F = Eq$.
- *Presently* only the intensity of the force field and the potential are defined, and then only as how much is diverged from the actual spatial EM field or potential by a unit point *static* charge assumed at each and every point.

Correcting an erroneous interpretation of the *first* law of thermodynamics (4)

- At best, the present calculation of "the field" is only an "indication" of the point intensity of field and potential, for interaction with *static* charges. It's actually a statement on *reaction cross section of the unit point static charge.*
- If the charge self-resonates at the frequency of the input energy, its interception (divergence) of energy increases, resulting in a different "field intensity defined" for the same "field". E.g., for the same amount of charge, the resonance case generates 18 times as much output energy from the same input as does the static charge case,* with no change in the *operator's* input. I.e., COP = 18.
- Presently no one calculates "the field" or "the potential" *a priori*, but only its "dependent point intensity" as determined for *static* charges.

* See Craig F. Bohren, "How can a particle absorb more than the light incident on it?" Am. J. Phys. 51(4), Apr. 1983, p. 323-327.

Contrary to popular opinion, perpetual motion is permissible and well-known

- Newton's first law of motion is *the law of perpetual motion*.
  - An object placed in motion remains in that state of motion in a straight line, unless changed by an external agent.
  - With no force interruption, the state of motion is *perpetual.*
  - The object needs no input of energy, and it does no work.
- Solid state physics students *do* perpetual motion experiments.
  - Once initiated in a superconducting ring, a persistent superconducting current will continue perpetually.
  - Careful measurements and proper statistical analysis indicates a half-life expectancy of some 10^{23} years.* The accepted age of the universe is about 1.4 x 10^{10} years.
  - The persistent ring current will last a great many orders of magnitude longer than the present universe has existed.

* Charles Kittel, Introduction to Solid State Physics, Seventh Edition, Wiley, New York, 1996, p. 359.

Common objection to COP > 1.0 power system as a forbidden perpetual motion machine is nonsense (1)

- Max Planck stated the erroneous objection this way:
  "It is in no way possible, either by mechanical, thermal, chemical, or other devices, to obtain perpetual motion, i.e., it is impossible to construct an engine which will work in a cycle and produce continuous work, or kinetic energy, from nothing."
- The statement contains two premises, one false and one true. It erroneously equates the two, and assumes that this proves the false premise. That is a simple *logical non sequitur*.
  - False premise: *Perpetual motion is impossible*. Falsified by Newton's 1st law and today's persistent super-conducting experiments in solid state physics labs.
  - True premise: *No engine can do continuous work in a cycle without energy input*.
  - The "i.e., ..." erroneously equates the two as the same thing. That is a non sequitur.

Common objection to COP > 1.0 power system as a forbidden perpetual motion machine is nonsense (2)

- That this rather inane non sequitur has not been noticed for more than 100 years of using it for "knee-jerk" condemnation is inexplicable.
- Perpetual motion has nothing to do with performing work or receiving extra energy input. It is simply Newton's first law of motion.
- Perpetual motion is *experimentally* proven. From several sources, one can even purchase a standard kit and perform a successful perpetual motion "persistent superconducting current" experiments oneself.
- Can have a perpetual motion machine doing work, if:
  - Necessary energy is received (as from the vacuum).
  - On the average the system is a NESS system.
- E.g., one can have a "self-powering" EM power system.

Permissible working perpetual motion machines with continuous energy input (NESS systems) can power our systems

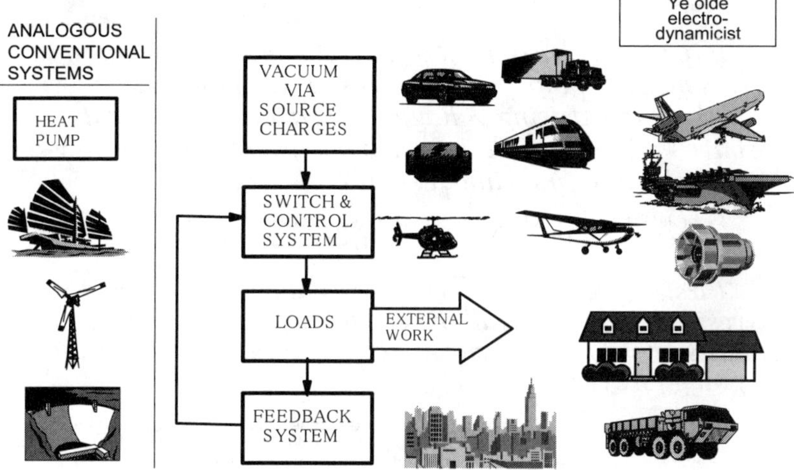

An interesting fact: The entire scientific community can be wrong for more than a hundred years

- In science model errors, often simple foundations are wrong, rather than the complicated mathematics.
- E.g., for more than 100 years the scientific community has automatically condemned "perpetual motion" and "dirty old forbidden perpetual motion machines".
- It seems that no one has previously subjected the typical nonsense statement to a simple sophomore logic analysis.
- Indeed, it continues to be used to denigrate EFTV, cold fusion, and other researchers and to slander them.
- To borrow a phrase from Nikola Tesla:

  "... [Eventually] it will be recognized as one of the most remarkable and inexplicable aberrations of the scientific mind which has ever been recorded in history."

Some true statements on perpetual motion (1)

"If a current is set up in a superconductor with multiply connected topology, e.g. a torus, it will flow forever without any driving voltage. (In practice experiments have been performed in which persistent currents flow for several years without signs of degrading)."

Peter J. Hirschfield, U. of Florida, Physics Teaching Notes, Chapter 5, Superconductivity, http://www.phys.ufl.edu/~pjh/teaching/phz7427/7427notes/ch5.pdf.

"First, there is no electrical resistance. There's no resistance because all the electrons are collectively in the same state. ... A current once started, just keeps on going forever."

Richard P. Feynman, Robert B. Leighton, and Matthew Sands, The Feynman Lectures on Physics, Vol. III, Quantum Mechanics, Addison-Wesley, Third Printing 1966, p. 21-08. Feynman was speaking about persistent superconducting currents.

Some true statements on perpetual motion (2)

- Roy* sums it up elegantly:

"It follows from Newton's laws that an isolated system in motion, on which no [net] force or torque is acting, exhibits precisely perpetual motion of the second kind. An example of perpetual motion of the second type is the orbiting of electrons around the atomic nucleus. ... perpetual motion of the second type is common on atomic and celestial scale; however, such a motion is not common in everyday life. The best known example of perpetual motion in everyday life is superconductivity, in which a current circulates ceaselessly in a wire loop without a battery."

* Bimalendu N. Roy, Fundamentals of Classical and Statistical Thermodynamics, Wiley, Chichester, 2002, p. 59.

The march of time in physics

- Time is not observable. All observables are spatial, not spatiotemporal. No observable exists in time, *a priori*.
- "Observation" is a $\partial/\partial t$ process imposed upon an ongoing 4-space (spatiotemporal) process. Dimensionally it gives a momentary snapshot:

$$\partial/\partial t\,(L^3T) = L^3$$

- An *observation* yields a *frozen 3-space snapshot at one instant in time*. The next instant, another snapshot is taken. And so on. We "see" movement of each piece of the "observed" by (a) its continual observation at high speed, yielding snapshots much like the frames of a movie, and (b) recall and comparison.
- The photon interactions with an observable generate its "march through time". Continuity is achieved by recall and comparison in the mind of the observer.

More sweeping changes to physics (1)

- What we refer to as "propagation of energy through 3-space" is in fact successive 3-positioning of a circulation of energy in 4-space, between the time-axis and a 3-space point and back, or vice versa.
- This latter 4-circulation is what we refer to as "charge". Hence negative charge is energy circulation from T to L^3, and positive charge is energy circulation from L^3 to T.

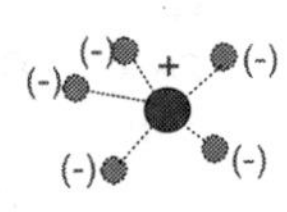

- *Charge conservation* is the complete 4-circulation. Due to vacuum polarization, any classical charge is a dipolarity, hence involves both energy flows.

More sweeping changes to physics (2)

- Zero charge may be either *absence* of circulation, or the *presence* of the same magnitude of equal and opposite circulations. The first is stress-free, while the second produces a stress 4-potential, a change in energy density of the vacuum, and a curvature in spacetime.
- Propagation of EM energy through 3-space actually is the successive positioning of the 4-circulation — the *propagation* of a 4-circulation.
- This explains Silverman's comment that *"we really do not know what charge is ...",* by revealing at least a view of *how charge can be modeled.*

More sweeping changes to physics (3)

- "Massless cause", "massless field", "massless potential", "curved spacetime", "exchange of virtual particles" are all slightly different aspects of the same precursor: *massless cause* (massless potentials and fields in space and changes of VPF in vacuum).
- All apply to the unobservable condition that exists prior to observation (observation occurs only after interaction of precursor with mass).
- Observation of the process whereby an ongoing massless cause is interacting with a previously observed "effect" to change it, is a $\partial/\partial t$ operation induced upon that LLLT process to produce a new "frozen LLL snapshot" (observed effect).

More sweeping changes to physics (4)

- The ∂/∂t operation induced upon ongoing 4-interaction between a 4-cause and a previous 3-effect, yields the newly observed "form" of the observable, which provides the "change" to it (when compared in memory to previous effect observed).
- The best physics and engineering is to consider a set of such 4-causes — that will interact with or are interacting with a system — to be a nonmaterial, force-free precursor "engine". Directly engineering and using such "precursor engines" is a new and more fundamental approach. It is in fact that of FSB/KGB energetics.
- Precursor engineering (energetics) is *a priori* an engineering grand unified field theory approach.

Peculiar note on Dirac "hole" thermodynamics

- In thermodynamics, several areas are already known and accepted to violate the second law. One of these is sharp gradients, about which *"Not much is known either experimentally or theoretically"*.*
- A very sharp gradient in energy density across a region of vacuum lifts electrons out of the Dirac Sea, leaving behind unfilled holes. Some of these holes can and do persist, not meeting electrons to "eat", refill and vanish.
- The hole (negative mass energy, negative energy EM fields and potentials) *is not* a positron (positive mass energy, positive energy EM fields and potentials).

* p. 459, Dilip Kondepudi and Ilya Prigogine, Modern Thermodynamics: From Heat Engines to Dissipative Structures, Wiley, Chichester, 1998, Reprinted with corrections Oct. 1999.

Note on dark energy and dark matter problems (1)

- Dark energy is just *negative EM energy*. Here's how.
- Persisting or temporarily persisting Dirac sea holes are *negative mass-energy* "source charges".
- While existing, as a source charge a hole produces negative energy EM fields and potentials, spreading at light speed from the moment of hole formation by sudden removal of its filling electron by sharp energy gradients.
- When the long-ignored Heaviside curled energy flow component is also considered, the negative energy field energies involved are perhaps a trillion times greater than conventionally can be estimated via Poynting theory.
- Nonsignificant antigravitational effects are thus produced, by negative mass of holes and by negative energy fields.

Note on dark energy and dark matter problems (2)

- A fraction of Dirac holes produced in violent gradient processes in the galaxy do persist. They are repelled away from the masses producing them, and stream toward the outside of the galaxy.
- As they exit the galaxy, they slow due to increasing repulsion from the matter of distant galaxies. Thus with passage of time there is a slowly increasing "pooling" of Dirac holes (dark matter) outside each galaxy.
- These increasing pools gravitationally repel the originating galactic matter, reducing that matter's escape due to velocity outward and centrifugal forces.
- The pools also gravitationally repel the matter of distant galaxies, so that a gradually increasing expansion acceleration occurs.

Note on dark energy and dark matter problems (3)

- In the dark matter problem, the G-forces produced on the galactic arms to hold them together have been assumed to be *positive gravity* forces from hidden (dark) mass inside the galaxy and arms.
- Instead, it is proposed that those forces are in fact *antigravity* forces — from accumulations of Dirac holes (negative mass matter) just outside the galaxy — acting back on the galactic matter to repel it.
- As source charges, these persisting holes continue to radiate negative energy photons, producing and sustaining negative energy EM fields and potentials. Those contain the missing "dark energy" of the cosmos.
- Dark energy is just the negative energy fields and flows from dark (negative) matter as source charges.

Dark (negative) energy and dark matter: Galaxy integrity and expansion of the universe (1)

- A sudden impulsive and very strong energy gradient across a region of vacuum lifts electrons from the Dirac sea, flowing in one direction, and leaving behind unfilled negative energy Dirac holes flowing in the opposite direction. The flowing Dirac holes are currents of "dark matter" (matter having negative mass energy).
- The negative-energy hole *is not* a positive-energy positron, contrary to the prevailing interpretation. A positron has *positive* mass-energy and *positive energy* fields; a hole has *negative* mass-energy and *negative energy* fields. Equating the two has always been a logical non sequitur.
- As a source charge, the persisting hole emits negative energy photons, establishing and continually replenishing its associated *negative energy* EM fields and potentials, spreading at light speed.
- Violent sharp-gradient processes in galactic matter continually produce such outward light-velocity streams of negative energy fields followed by much slower streams of dark matter (holes).

Dark (negative) energy and dark matter: Galaxy integrity and expansion of the universe (2)

- The outward-spreading dark (negative mass) matter streams are slightly repelled gravitationally by positive-energy masses in the galaxy. Thus they "navigate" slightly through a galaxy, avoiding matter except for those electrons forcefully meeting them and recombining without radiation.
- The dark matter tends to slow down outside the galaxy and exhibit dynamic cumulation (nonequilibrium steady state) outside, due to gravitational back-repulsion from the cumulations of negative energy outside other galaxies, etc.
- The cumulation of this dark matter around and outside each galaxy repels positive matter back into its own galaxy and also repels positive matter in all distant galaxies.
- Onward-racing negative energy fields with their Heaviside energy components contribute substantially to the effect.

Dark (negative) energy and dark matter: Galaxy integrity and expansion of the universe (3)

- Considering the long-neglected giant *Heaviside energy component* of this negative EM energy flow and dark matter flow, its cumulation outside the galaxy far exceeds conventional calculations and leads to significant gravitational repulsion effects.
- From the collection/cumulation outside the galactic masses, the gravitational repulsion *back into* the galactic matter (e.g., back into a spiral galaxy's arms from the altered space surrounding each arm) holds the galactic matter together, preventing it from flying apart due to centrifugal forces overriding the normal positve gravitational attraction.
- This is largely the phenomenology erroneously labelled the "dark matter" effect. It is not an attraction from inside, but a repulsion from outside.

Dark (negative) energy and dark matter: Galaxy integrity and expansion of the universe (4)

- Continuing accumulation of dark matter and negative energy concentrations outside each galaxy results in the increasing mutual repulsion of galaxies. At some point in the past, with the increasing Dirac hole cumulations the increasing mutual gravitational repulsion of the galaxies caught up with and exceeded their gravitational attraction, and it continues to slowly increase, further exceeding it.
- This reversed the *deceleration* of the expansion and generated and continues to generate the *acceleration* of the expanding universe, by gradually increasing the mutual departing intervelocities of the galaxies.
- Originally positive G attraction between galaxies was slowing universe expansion. Finally the cumulating antigravity caught up and surpassed it. So now the expansion of our universe is increasingly accelerated.

Dark (negative) energy and dark matter: Galaxy integrity and expansion of the universe (5)

- Negative energy and dark matter can be investigated on the bench, using Bedini techniques* and other techniques used by Sweet.**
- Together with its Heaviside component, dark energy from the dark matter produces significant antigravity effects** and many other effects.

---

* John C. Bedini and Thomas E. Bearden, "Radiant Potential Energy Charger", U.S. Provisional Patent Application ER 677978580 US, 2004. A formal patent application is in preparation.

** Floyd Sweet and T. E. Bearden, "Utilizing Scalar Electromagnetics to Tap Vacuum Energy," Proc. 26th Intersoc. Energy Conv. Eng. Conf. (IECEC '91), Boston, Massachusetts, 1991, p. 370-375. A highly successful antigravity experiment was performed, with the VTA unit losing 90% of its weight on the bench. Results are reported in the paper.

Dark (negative) energy and dark matter: Producing anomalous Pioneer drag

- In a galaxy, violent gradient processes are continually occurring, continually producing dark matter and flows of negative energy slowly cumulating outside the galaxy. Dark matter accumulations outside produce more antigravity fields *back into* the galaxy, acting on galactic matter, than does the thinner concentration in the flow inside the galaxy. A similar effect surrounds the sun.

NASA's Pioneer spacecraft

- A spacecraft moving away from our sun towards the outside of the galaxy, experiences a "net drag" force upon it, due to the excess G-repulsion from the outside dark matter accumulation. This accounts for the anomalous drag experienced by NASA's Pioneer 10 spacecraft and some others.

Some aspects of Bedini's important negative energy investigations (1)

- EM energy flow occurs in two main forms:
  - Divergent (*positive*) energy flow, where the energy is always trying to diverge out of its flow direction through space and escape back to the environment.
  - Convergent (*negative*) energy flow, where excess negative energy from the vacuum environment is trying to enter and increase the ongoing flow of negative energy.
- "*Conductivity*" means the ability of the conductor to "hold in" the positive energy on its flow path through space, and to "hold back" excess negative energy from the vacuum attempting to freely flow into and increase a negative flow of energy through space.
- "*Impedance*" is reduction of conductivity (ability to hold in or hold back the change in energy flow).

Some aspects of Bedini's important negative energy investigations (2)

- Positive energy flow encountering an impedance will "lose some energy" diverged from the energy flow path. Less energy flow along the path leaves the impedance than entered it.
- Negative energy flow encountering an impedance will "gain some negative energy" converged in from the environment. More negative energy flow along the path leaves the impedance than entered it.
- Negative energy flow passing through a series of impedances can undergo tremendous gain, freely receiving excess negative energy from its environment.

Some aspects of Bedini's important negative energy investigations (3)

- Negative energy flow passing through a series of impedances freely receives excess negative energy from its environment via each "impedance".
- If loads are designed for negative energy, impedance of load provides an additional energy flow gain.
- Such negative energy flows are nonequilibrium systems, freely receiving excess negative energy from their environment.
- In theory, a flashlight battery producing a little negative energy could be used in a circuitry of special iterative series impedances to power a large city.
- The vast excess EM energy gathered would freely come from the vacuum environment.

Conversion of negative EM energy to positive EM energy

- Converting negative energy to positive energy is simple.
  - Charge a capacitor (or battery) with the negative energy, by connecting the battery "backwards" and pulsing it with negative energy pulses.
  - The battery charges rapidly, and also undergoes appreciable rejuvenation inside.
  - The battery charges with much less "operator-input" energy, because the environment freely inserts <u>*additional*</u> negative energy via the impedance of the battery.
  - Switch away the rapidly charged battery into another circuit, connecting it normally.
  - <u>*The battery discharges more "normal positive EM energy" into the loads and losses of that circuit than the operator furnished to do the switching and controlling*</u>.
- Scientific study and development of negative energy loads and systems can rapidly solve the energy crisis forever.

Einstein's admonition

"... the scientist makes use of a whole arsenal of concepts which he imbibed practically with his mother's milk; and seldom if ever is he aware of the eternally problematic character of his concepts. He uses this conceptual material, or, speaking more exactly, these conceptual tools of thought, as something obviously, immutably given; something having an objective value of truth which is hardly even, and in any case not seriously, to be doubted. ...in the interests of science it is necessary over and over again to engage in the critique of these fundamental concepts, in order that we may not unconsciously be ruled by them."

(SLIDE BRIEFING CONTINUED)

PART I: APPENDIX; Vulnerability of Network-Centric Warfare

Appendix 2: *Example*: Vulnerability of Network-Centric Warfare

Nullified by asymmetric energetics warfare

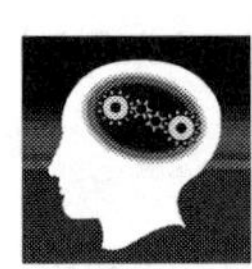

Basic idea in Network-Centric Warfare NCW)

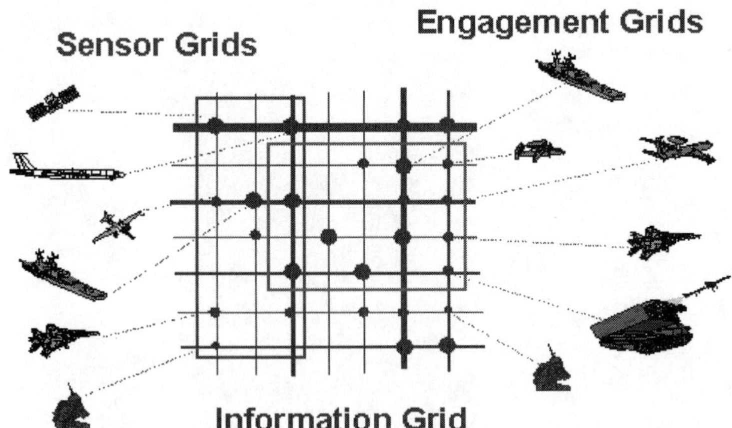

Network-centric warfare (NCW) in part of the theater

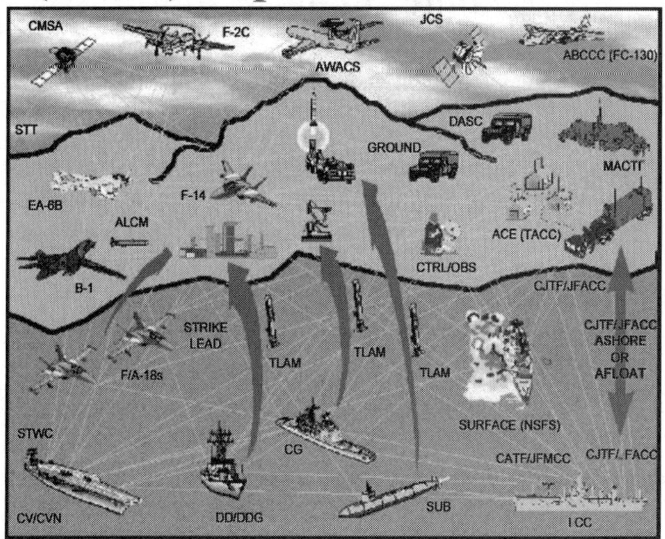

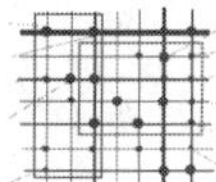

Network-centric warfare

- A developing U.S. style of warfare based on force transformation and total network implementation on the battlefield, to unify and coordinate resources, forces, and actions.
- Particularly directed toward interoperability.
- Ensures all facets are "playing from same sheet of music".
- Shifts sources of power to information-based activities.
- Commanders effectively direct the various elements of the action, in real-time or near-real-time.
- Replaces the older notion of "orchestrating the battlefield", but goes much further by accenting interoperability, communications, and architecture.

* E.g., see Max Boot, "The Struggle to Transform the Military," Foreign Affairs, Mar./Apr. 2005, p. 103-118. Also see briefing on http://sunset.usc.edu/gsaw/gsaw2005/s8/slater.pdf

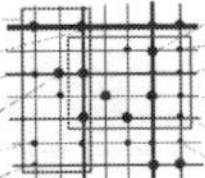

Network-centric warfare includes many good things

- Continuous adaptive acquisition
- Robust operational prototyping
- Capability in the hands of people as possible, then issue the next item, then the next, constantly growing and changing
- Better than forecasting a need 15 to 25 years hence, freezing design then making it fit into the reality that emerges
- Work-in strategy threats and technology first, which yield concept development
- Balanced approach between creativity and implementation
- Train for the known but educate for the unknown
- Based on mission capabilities package — the sum total of that which is necessary to perform a task

---

Largely taken from presentations by Admiral Arthur Cebrowski, Vice Admiral USN (Ret.), who is Director of Office of Force Transformation.

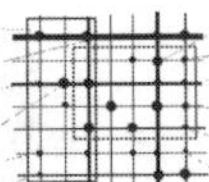

Basic intent of NCW (1)

(Also see following slide)

- Time delays out of <u>*Observe, Orient, Decide, Act*</u> (OODA) loop.*
- Immediacy of NCW information permits accomplishing OODA cycle on a smooth, nearly continuous basis at all levels of command, for "empowered self-synchronized" operations.
- With training, immediacy of near-total information available at all levels should further accelerate the semi-independent operations for "much greater speed of command."
- Permits decentralizing or flattening the command structure, taking the control function down to the lowest practicable level of command and shortening the response cycle.

---

* Col. John Boyd's famous loop, widely known in strategy circles.

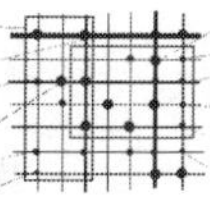

Basic intent of NCW (2)

Self-Synchronization and the Speed of Command

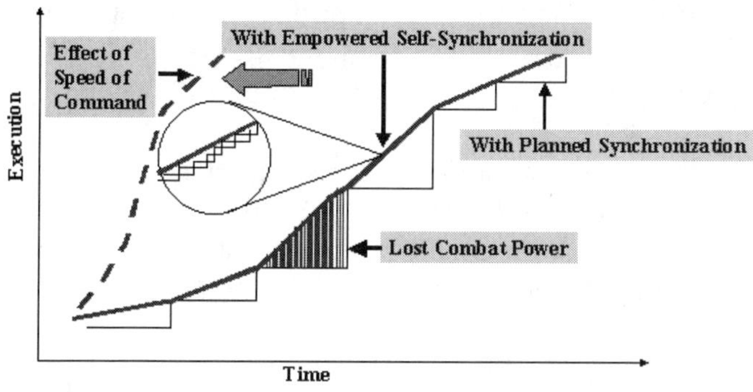

New Sciences and Warfare
VADM A.K. Cebrowski 9/21/98

The asymmetric response

Marshall Andrei Grechko

- *"History knows many examples how the employment of a new weapon initially gave considerable success because the enemy, caught unawares and not knowing the combat capabilities of this weapon, was for some time incapable of effective counteraction."*

  (V. Yo. Savkin, The Basic Principles of Operational Art and Tactics, Moscow, 1972.)

- *"Of particular importance is basic research aimed at discovering still unknown attributes of matter, phenomena, and the laws of nature, and developing new methods for their study and use to reinforce the state's defense capability."*

  (Marshall Andrei Grechko)

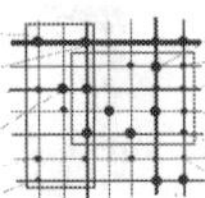

Network-centric warfare: Assumptions defeated

- NCW assumes GO — Group Order and coherence — in operation of equipment, networks, personnel, and units. That is an *already-countered* assumption.
- When the battlefield opponent fits its *implicit assumptions*, NCW is superb and devastatingly successful.
- When the assumptions are asymmetrically violated via *energetics* means, NCW is an instant disaster.
  - Rapid destruction of communications and order.
  - Rapid defeat of disarrayed *NO GO* U.S. forces in short order.
  - Against an opponent (such as FSB/KGB, Yakuza, or China) armed with energetics weapons, NCW is already rendered obsolete.

Incident in World War I

- In WW I, in one incident commanders produced a rigid plan of attack, massive, and down to the finest details.
- Predicted superb success for such a superb plan.
- The attack was a total disaster. The Germans violated the programming assumptions and chewed up the operation.
- **Lessons to be learned:**
  - A strategy may work against an opponent fitting its implicit assumptions.
  - It fails miserably against a foe who can and does deviate asymmetrically from the implicit assumptions.
- **Remark**: Network-centric war worked decisively against a highly inferior Iraqi military, but proved nothing at all for asymmetric warfare in a modern *energetics weapons* arena.

Programmed forces: Gettysburg

- At the battle of Gettysburg, forces reacted with programmed or stereotyped responses.
  - Both Blue and Gray forces followed their programs.
  - Commanders were powerless to change their own tactics.
  - Confederate cavalry under Stuart was absent the first day.
  - 17,000 men fell in one day.
- It turned the tide of the Civil War against the Southerners.
- We speak of "BlueGray War" or "BlueGray conflict" as maximum killing conflict, in a force that is set (programmed) against itself (in civil warfare), *particularly by energetics means*.
- *Conversion* enables the ultimate "reprogramming by higher power" and thus the ultimate BlueGray war.

Part of a song* about the Battle of Gettysburg

"Seventeen thousand fell today;
The cannons roared and blew them all away.
Blood ran red from Blue and Gray,
And seventeen thousand fell today."

"Broken bodies lying all around,
Mould'rin' in the sun and on the ground.
Pigs and vultures came from far away —
Lawd, they had a feast laid out today."

---

* "Seventeen Thousand Fell Today," T. E. Bearden, 1999. And yes, the pigs and vultures did feast outside Gettysburg on the great number of fallen bodies, both blue and gray.

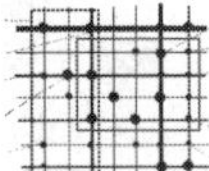

GO assumptions in NCW become NO-GO in reality

- Against energetics weapons, stability of C4 (command, control, communications, and computer systems) quickly vanishes.
    - Signals fail.
    - Facilities fail.
    - Equipment fails.
    - Personnel fail.

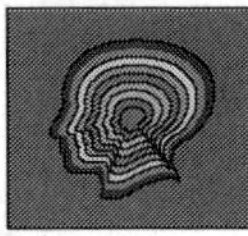

- Energetics warfare on all three fronts (energetic, bioenergetic, psychoenergetic) completely erases stability in all four areas.
- Initiates immediate and ultimate BlueGray internal conflict.
- *Against a modern energetics warfare foe, the network-centric force without energetics nearly instantly succumbs*.
- With psychoenergetics, GO becomes NO-GO in the targeted soldiers, regardless of how large the group. Group ordering instantly vanishes, and utter chaos is explosively induced.

For a world army of energetics robots

- After sunk costs on the front end, replica precursor robots cost about one cent each.
- A billion "carrier" robots — each with a thousand "secondary" weapons robots aboard — is about 1.001 trillion weapons costing a bit more than $10 billion.
- This is sufficient to attack and destroy all modern nations, peoples, installations, weapons systems, etc. many times over. Nothing is immune or protected.
- Even $1 billion (and time to reproduce all those robots and expense to get control of them) is sufficient to allow destroying all one's enemies with their weapons systems, populations, facilities, and installations.

Some example scenarios

- **New National Emergency scenario.**
- **Chinese scenario.**
- **Terrorist scenario.**
- **Gradual defeat scenario.**
- **Dissolution of society scenario.**
- **Sudden defeat scenario.**
- **Massive sleeper conversion scenario.**

Scenario 1: Substantial new national emergency arises

- U.S. President is whisked on board Air Force One, complete with entourage.
- Key congressmen are whisked to secret underground facility.
- Alert forces react, patrolling skies, border, oceans, cities, etc.
- Command and control centers on full alert and in emergency operations.
- ICBM, bomber, and nuclear subs ready and/or deployed.

Kill of President and Vice President:

Use of ultimate "throwaway" assassins (1)

- Converted staff and security personnel on board the Presidential aircraft have personas triggered by implanted, waiting robots.
- BlueGray conflict erupts on AF-1. Converts kill president and his party.
- Open gun battle between security people. Converts kill others, who are completely outnumbered.
- Converted pilots calmly fly AF-1 into the side of a mountain, turning it into a fireball and dying with it.

WHEN REAGAN SHOT

Kill of President and Vice President:

Use of ultimate "throwaway" assassins (2)

- At same time, other conversions are triggered in separate facilities of the Vice President. Converts destroy facilities and kill the VP.
- Simultaneously, triggered conversions in congressional facilities occur. Converts methodically kill all personnel and congressmen, then destroy facility with themselves in it.
- Facilities and personnel for Joint Chiefs, Supreme Court, State Department, etc. are destroyed in similar fashion.
- Very quickly the U.S. is devoid of most of the leaders of its government. Process can continue further downward.

Sleeper conversions triggered and key personnel destroyed

- Key personnel *converted;* energetics robots begin destroying equipments, C4, weapons, patrol vehicles, etc.
- BlueGray warfare erupts throughout security forces as equipment fails or is destroyed.
- Converted and nonconverted personnel in security forces "eat themselves" by killing each other off.
- Command and control centers "eat themselves" and fail.
- Entire strategic force has bombers, weapons, nuclear subs, carriers, etc. destroyed by robots. Converted personnel attack and "eat each other" and destroy own friendly facilities by BlueGray conflict.
- Entire U.S. is utterly destroyed, as robots finish mop-up.

Additional actions

- Massive robot attacks and conversions unleashed on cities, civilian populace, facilities such as power stations and refineries, ports, etc.
- Civilian populace in full BlueGray war because of extensive conversions. Populace consumes itself.
- Power stations, refineries, port facilities, bridges, major buildings and facilities destroyed.
- America becomes a smoking ruins.
- Anew, *"Pigs and vultures came from far away,*
  Lawd, they had a feast laid out today."

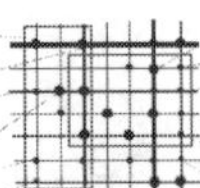

End result: GO to NO-GO

- System *stability and survival assumptions* of network-centric warfare are totally negated.
- Without energetics defense, network-centric warfare forces can be swiftly and easily destroyed in the era of asymmetric energetics weapons.
- Nuclear weapons, nuclear propulsion systems, and nuclear power systems are easily and quickly dudded by QP weapons.
- Strategic capability of the United States is very rapidly destroyed.

Scenario 2: Chinese scenario (1)

- China prepares both ways: to accommodate, and to strike and destroy.
- FSB/KGB inserts mass "sleeper" personas in all key Chinese officials.
- Cooperation (or conversions) in Chinese at various levels, and in targeted nation's (U.S.'s) port and security personnel.
- Chinese insert numerous shipping containers filled with nuclear weapons and other WMD, into United States. Converts in U.S. ports and in officialdom assist overtly.
- 50 to 100 Chinese nukes inserted in U.S. cities.

Scenario 2: Chinese scenario (2)

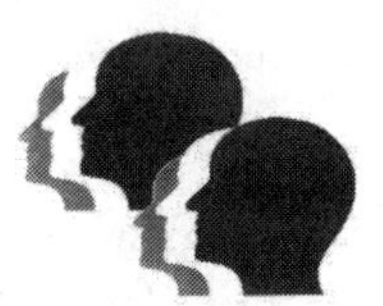

- Inserted nukes are detonated, in the dirt, destroying major U.S. cities; 150 million U.S. casualties.
- BlueGray war erupts; Yakuza finishes off U.S.
- FSB/KGB then promptly retriggers massive sleeper Chinese conversions.

- Violent BlueGray war erupts in China and China destroys itself very quickly.
- FSB/KGB emerges winner.
- If deemed necessary, sleeper conversions in Yakuza destroy Yakuza also.

Scenario 3: Terrorist scenario

- Organized terrorist leaders undergo conversions, as do selected ancillary U.S. personnel etc.
- Converts in several nations provide nuclear weapons to terrorists (such as Al Qaeda).
- Containers filled with nuclear weapons openly inserted in U.S.; U.S. converts assist directly.
- 50 to 100 nukes inserted into U.S. cities, also other WMD.
- Terrorist converts detonate nukes, destroying cities and producing 150 million casualties.

- U.S. BlueGray war erupts due to internal conversions. Yakuza finishes off the U.S.
- FSB/KGB retriggers terrorist converts.
- BlueGray war consumes terrorist organizations, Arab nations, etc. which eat themselves.

Hypothetical conversation

- Q: *"How many dedicated hard-core terrorists would you like inserted in the United States?"*

- A: *"One million."*
- Response: *"Piece of cake. From a distance we will put sleeper conversion personas in a million carefully chosen Americans (young, strong, active, intelligent, knowledgeable) and suddenly switch them."*
- Q: *"Will they be dedicated, and will they be willing martyrs?"*
- A: *"Far better than any you presently have available for insertion. They will also be unrecognizable prior to the conversion switching."*

Scenario 4: Gradual defeat scenario (1)

- Conversions in selected elements of U.S. society, with sophisticated switchable personas.
- Polarizations — disagreements, oppositions, fanaticism — intensify slowly in polarized society.
- Courts, law bodies, law enforcement, government levels gradually split into multiple warring factions.

- Entropy of society grows steadily and without bounds, spurred by slow increase in converts who keep intensifying polarizations. There is chaos in the streets.
- Entropy and conflict increase to where paralysis begins.

Scenario 4: Gradual defeat scenario (2)

- Economy collapses, culture collapses.
- Will, emotional intent, and ability to get things done wither away and die. U.S. society grinds to a halt, consumes itself.

- U.S. is "defeated" mentally and emotionally, fading into entropy (anarchy) with direct self-killing and intense BlueGray war. GO becomes NO-GO.
- Yakuza scalar interferometers augment the chaos and destruction, assisting in causing catastrophic U.S. economic collapse.
- Then Yakuza methodically completes total destruction of the U.S. with its scalar interferometers.

Scenario 5: Dissolution of society (1)

- Switchable persona conversions in selected leaders of U.S. activist groups, influential groups, media groups.
- Some gradually switched into conversion; deepen disinformation and activist programs.
- News media, government agencies, activists perverted.
- U.S. groups set against other U.S. groups, with slowly increasing fanaticism.
- BlueGray war between groups starts and increases steadily in intensity. Society begins to consume itself.

Scenario 5: Dissolution of society (2) 

- Purposeful and concerted peaceful actions mostly cease; nation and culture come unglued as senseless violence begins and escalates.
- National culture unravels, polarized opposing groups are "eating each other". Dissolution of the nation, of the culture, and of "normal" GO efforts such as by populace, military forces, police forces.
- Electrical power system destroyed; economy violently collapses.

- Robots and Yakuza "finish off" remainder of populace and society.

Scenario 6: Sudden defeat

- Massive converts triggered by "switching-in" pre-installed "sleeper" persona conversions.
- Extensive BlueGray warfare erupts in fierce intensity.
  - Precursor robots simultaneously attack massively, destroy installations, cities, facilities, power systems, weapons systems, command and control systems, etc.
  - Terrorists simultaneously unleash anthrax, smallpox, nukes, scalar interferometry.
- At same time, converted U.S. armed forces groups begin attacking American cities and each other with U.S. nukes, biowar, missiles, and field forces. U.S. military neutralized.
- Total BlueGray warfare without rhyme or reason rapidly consumes America, collapsing the populace and the entire nation. Then Yakuza methodically finishes off remnants.
- Can occur in two days or less; even in two hours.

Scenario 7: Massive sleeper conversion

- Pre-insert switchable sleeper personas in millions; all leaders.
  - Congress, president, cabinet, high officials, defense leaders.
  - Particularly saturate leadership of WMD retaliation and deterrence, military, police.
  - Particularly include Israel.

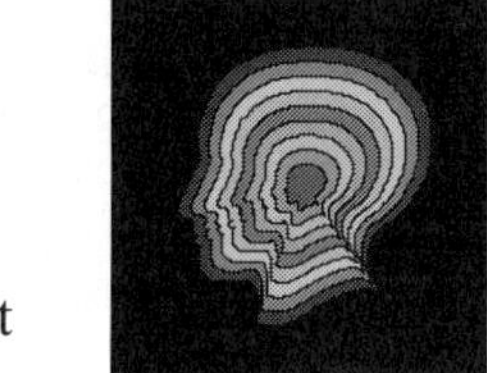

- Evoke "sudden destruction" scenario, just after switching-in all sleeper conversions.

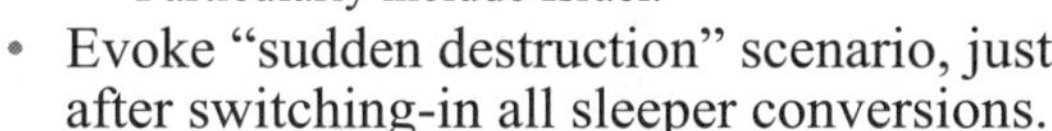

- Entire society quickly nullified as all sanity and ordering lost. President, Congress, pentagon hierarchy, all security hierarchies commit suicide or destroy each other and populace.
- Society destroys itself at same time that Yakuza, terrorists, and others are attacking and destroying any remaining vestiges.

Important conclusion

- Present scenario chosen seems to be a combination of the slow dissolution scenario and the massive sleeper scenario.
- Must have the massive sleeper scenario to nullify Israel, else FSB/KGB and Russia destroyed also.

- Assuming the combination:
  - The slow scenario for us lasts two years, peaking to destruction of U.S. centralized electric power system, refineries, etc. and with catastrophic economic collapse.
  - Just prior to that climax, sudden dramatic demise of Israel occurs via evoking the massive sleeper scenario upon that state.
  - Sleeper conversion personas already inserted in both the U.S. and Israel, perhaps UK and others, in massive numbers.
- The Yakuza will play the major protégé role in the internal overt destructive actions against the U.S.
- U.S. is unprepared to detect or defend against inserted sleeper conversion personas, triggerable at any time.

Remarks

- Hundreds of other scenarios exist and can be initiated. Limited only by imagination and ingenuity of planners and by availability of energetics weapons and capabilities.
- All concepts that "military organizations and systems keep functioning and viable as totally reliable forces" are in error from the outset by conversions, including U.S. network-centric warfare.

- Presently planned Army of the Future is neutralized and defeated by energetics warfare. It can be neutralized in the mind, body, equipment, signals, organizations, and functions modes simultaneously.
- One or more of these or similar scenarios are now in place and already slowly progressing in the operations phase of asymmetric war against us.
- By early 2007, operations phase is expected to destroy us.

Final urgent commentary (1)

- Russian energetics has secretly applied Einstein's admonition for more than a half century.
- Long ago worked the major "kinks" out of EM, GR, quantum theory, etc. and achieved a grand unified field theory (GUFT) that has been highly weaponized.
- Our immediate problem is to sufficiently catch up at the most desperate speed possible, particularly in psychoenergetics. Barrier is U.S. scientific community.
- Presently we cannot defend against the eerie *new* destruction threats rapidly converging upon us, at the same time different but equally destructive threats are converging or have reached total destruction capability.

Final urgent commentary (2)

- The greatest problems are our severe lags in our
  - Century of delay in correcting the terribly flawed CEM/EE model.
  - 60 year delay in psychoenergetics, massive sleeper conversion, and the eerie nonmaterial robots.
- Unless U.S. science is sharply awakened from its century-long slumber, we are essentially already defeated and finished.

Final urgent commentary (3)

- We presently have one slim chance.
- Given that Israel does somehow have defenses in psychoenergetics, then Israel may be able to save the U.S. again, as it did in the 1970s, in 1986, twice in 1997, and again at the end of 1999 and in the first part of 2001.

- If, on the other hand, Israel itself will be destroyed this time, then nothing we have will save us.
- In that case, the U.S. scientific community will have destroyed us.

Final urgent commentary (4)

- The darkest days in the history of our republic lie immediately ahead of us.
- Hopefully present, rapid Israeli developments (and others) may reduce or nullify the conversion threat.

- If we are to survive, we shall need the most strenuous and rapid effort in our history, *now*.
- In times of grave peril, Americans have always rallied to national liberty and survival. We must do so again.
- God bless and keep America! We pray for its survival.

THE END

References

i. T. E. Bearden, Fer-de-Lance, Cheniere Press, Santa Barbara, 2nd edition, 2003, gives a reasonably good overview of the Russian/Soviet/KGB/FSB development and use of what is called *energetics weapons*. These are superweapons using a new approach for using *force-free fields in space* prior to their interaction with matter to form *force fields in matter*. We have also called this new approach "precursor engineering", in order to tie it to conventional Western physics as an extension thereof, and hopefully to convince Western scientific leaders to correct the seriously flawed and horribly archaic old classical EM and electrical engineering models taught today in Western universities.

The standard CEM/EE model is so fouled with known major errors that it has become a reflection on the lack of scientific ethics at the highest levels of our scientific community. Our scientific leaders continue to advocate and promulgate such erroneous models, and are directly contributing to the imminent destruction of our nation and people. They continue to ignore the serious flaws of the models as pointed out by imminent scientists such as Feynman, Wheeler, Bunge, Lindsay, Margenau and many others. This cannot be simply shrugged aside and ignored with impunity, if we wish to survive.

Because of its long inaction on this problem, the U.S. scientific community is directly responsible for the presently escalating energy crisis, and it is also partially responsible for the high probability of success of the present 3-year terrorist strategic program leading to our total destruction. It is further responsible for our having no effective "amplified EM anti-engines" treatment for mass casualties, so that millions of stricken Americans will die who did not have to, had the scientific community changed its ostrich position with its head firmly in the sand.

Any high level scientific organization or group involved in electromagnetics that is *not* discussing and working on the source charge problem — how a charge continuously pours out observable photons in all directions, without any *observable* energy input — is a scientific organization or group that is participating actively though unwittingly in the coming destruction of our nation and our people.

By the same token, any high level intelligence agency not addressing this seemingly innocuous "archaic CEM/EE model" problem will have no knowledge of the KGB energetics weapons and capabilities. Any such intelligence group is simply failing to perform its designated scientific intelligence mission. Sadly, almost all our present high level intelligence agencies have hardly even heard of the problem; most don't even care that it exists. They certainly have not considered such "foundations issues" to involve winning or losing the asymmetric terrorist war against us. Consequently, that war now enfolds and coordinates the international terrorists also, including the hidden key role being played by the Yakuza.

ii. T. E. Bearden, "Precursor Engineering: Directly Altering Physical Reality," Explore, 13(2), 2004, p. 50-60 somewhat crudely describes the basis for energetics, which the Soviets developed as a science of secret superweapons long since developed, deployed, and clandestinely being utilized against us.

iii. See T. E. Bearden, "Soviet Weather Engineering over North America," 1-hour videotape, 1985 for documentation of some of the early Soviet weather engineering operations, with pictures of the specific cloud signatures etc. Or just check the news regarding the recent 2004 Caribbean hurricanes being ginned up one after the other, strongly controlled, and strongly pounding the Southern and Eastern seaboards.

iv. See (a) M. W. Evans, P. K. Anastasovski, T. E. Bearden *et al.*, "On Whittaker's Representation of the Electromagnetic Entity in Vacuo, Part V: The Production of Transverse Fields and Energy by Scalar Interferometry," Journal of New Energy, 4(3), Special Issue, Winter 1999, p. 76-78. This paper gives a succinct mathematical development of scalar interferometry, using a higher electrodynamics with O(3) group symmetry. The original two papers by Whittaker are (b) E. T. Whittaker, "On the Partial Differential Equations of Mathematical Physics," Math. Ann., Vol. 57, 1903, p. 333-355, and (c) — "On an Expression of the Electromagnetic Field Due to Electrons by Means of Two Scalar Potential Functions," Proc. Lond. Math. Soc., Series 2, Vol. 1, 1904, p. 367-372. The two papers can be downloaded in PDF form from www.cheniere.org. The 1904 paper initiated that area of electrodynamics known as *superpotential theory*.

v. T. E. Bearden, Fer-de-Lance, 2nd Edition, Cheniere Press, 2003, with update, gives a rough time table for the development of these energetics weapons, the main types developed, and many actual test incidents for the weapons.

vi. T. E. Bearden, Gravitobiology, Second Edition, Cheniere Press, 2003 covers some of the biological aspects.

vii. T. E. Bearden and K. D. Moore, Conversion: A Major Strategic Threat, 2003, a 300+ page color slide briefing, covered the bizarre but highly successful mind-engineering (psychoenergetics) aspects. (It was an earlier version of the second part of this book). It was produced and delivered pro bono to a three-letter U.S. government agency. No open release of this document has been made by the authors. The document gives the basis for actual production of a living and functioning human mind in the laboratory, the insertion of that mind in another human being, and alternation (conversion) between the two minds to shift the behavior of the individual back and forth at will. Information on that subject is in the present paper.

What is demonstrated in a multiple personality case can be directly produced on the laboratory bench and in any human target anywhere on the planet.

An *operational fraction* of the mind can be used in conversion techniques also, without total mind switching but only with switching of that fraction involving a certain area or class of behavior.

Thus selected leaders in a societal polarization area, on both extreme ends, can be selectively and repeatedly "temporarily converted" to more extreme and dramatically different behavior and leadership, to push the polarization to greater intensity. That capability will be increasingly exercised over the next two years. The general background of the KGB development of this ability, actual dramatic human test incidents, and the science on which it is based are given in the present book. The application and role of conversion in the present 3-year coordinated terrorist program to destroy the United States is also included — briefly in Part I, and more extensively in Part II.

viii. On April 28, 1997 — even as Defense Secretary William Cohen was publicly confirming at a conference in Georgia that "terrorist" electromagnetic superweapons were being used to engineer the weather and climate, trigger volcanoes into eruption, and generate earthquakes — the moments were ticking away for a sudden massive Russian/KGB superweapon attack on May 1 that would have utterly destroyed the U.S. in two hours or less. Eerily, neither Dr. Cohen nor our U.S. intelligence agencies knew what was actually transpiring and coming down upon us, even as the Secretary spoke. Five days before the scheduled attack, we had sent a desperate fax to the prime minister of the supporting nation, pointing out a critical change that had to be made in their most powerful superweapon's command and control system. The change was quickly made and the Soviets were so notified. We waited in agony as the clock ticked until, with two days remaining before May 1 (the scheduled attack date) — and while Secretary of Defense Cohen was speaking in Georgia — that giant armada started standing down, and Armageddon was again avoided by a hair's breadth.

ix. Stanislav Lunev (with Ira Winkler), Through the Eyes of the Enemy, Regnery, Washington, 1998 gives a good accounting of the powerful part played by the Russian Mafia in the aftermath of the Soviet Union's economic collapse. Lunev is the highest ranking GRU defector. He is a former Colonel in the GRU, the military counterpart of the KGB. On p. 22-33, he summarizes the Spetznaz capabilities. On p. 22-27, he summarizes Spetznaz use of nuclear weapons already on American soil and in our cities. On p. 26, he gives some of the ways in which the Soviets easily brought nuclear weapons into the U.S. On p. 30 he confirms Russian seismic weapons. Use of EMP weapons is on p. 30-31. Use of very, very low frequency weapons to destroy the human brain, put people into a zombie-like state, and aid in brainwashing is confirmed on p. 31. Lunev was of course not allowed to detail the most highly relevant energetics weapons information. The book as published had to be cleared and approved by the U.S. Central Intelligence Agency, which would obviously not allow such details to be included.

x. For some details on the Yakuza, see (a) Brian Bremner, "The Samurai Meets the Godfather." Business Week, Jan. 29, 1996. On p. 47, Bremner gives a quick, half- page overview of the Yakuza in Japan. (b) — "How the Mob burned the Banks: The Yakuza is at the center of the $350 billion bad-loan scandal." Business Week, Jan. 29, 1996, p. 42-43, 46-47. This article summarizes the Yakuza involvement in, and cause of, the huge bad-loan banking scandal that occurred in Japan. This rip-off of the Japanese taxpayer enriched the Yakuza. See also (c) Michael Hirsh and Hideko Takayama, "Big Bang or Bust?" Newsweek, Sept. 1, 1997, p. 44-45. The Yakuza rip-off of the Japanese banking system may have been for as much as $600 billion. Our comment is that any high U.S. Japanese official seriously protesting the Yakuza involvement at the time, was in grave danger of being killed. Several were in fact assassinated by the Yakuza.

xi. Many details on the Aum Shinrikyo are given in Global Proliferation of Weapons of Mass Destruction, Part I, Senate Hearing 104-422, Hearings Before the Permanent Subcommittee on Investigations of the Committee on Governmental Affairs, U.S. Senate, Oct. 31 and Nov. 1, 1995. Sadly, that Subcommittee seems to have known nothing at all about the Soviet superweapons acquired by the Yakuza/Aum Shinrikyo, although that information certainly is known to certain U.S. Government agencies.

xii. We acknowledge the significant contributions of Australian geophysicist Harry Mason, who furnished this vital information and verified its authenticity.

xiii. See (a) Michael Hirsh and Hideko Takayama, "Big Bang or Bust?" Newsweek, Sept. 1, 1997, p. 44-45; (b) Brian Bremner, "How the Mob burned the Banks: The Yakuza is at the center of the $350 billion bad-loan scandal." Business Week, Jan. 29, 1996, p. 42-43, 46-47.

xiv. See (a) M. W. Evans, P. K. Anastasovski, T. E. Bearden *et al.*, "On Whittaker's Representation of the Electromagnetic Entity in Vacuo, Part V: The Production of Transverse Fields and Energy by Scalar Interferometry," Journal of New Energy, 4(3), Special Issue, Winter 1999, p. 76-78. This paper gives a succinct mathematical development of scalar interferometry, using a higher group symmetry electrodynamics. The original Whittaker papers giving the basis for scalar interferometry are listed in references {xxi, xxii} below.

xv. Graham H. Turbiville, Jr., Weapons Proliferation and Organized Crime: The Russian Military and Security Force Dimension, U.S. Air Force Academy, Institute for National Security Studies, Colorado Springs, CO, June 1996.

xvi. For conventional consideration of hurricane increases, see (a) Lennart Bengtsson, "Hurricane Threats," Science, Vol. 293, 20 July 2001 p. 440-441; (b) Stanley B. Goldenberg *et al.*, "The Recent Increase in Atlantic Hurricane Activity: Causes and Implications," Science, Vol. 293,
20 July 2001 p. 474-479. For a young weatherman who recognizes the direct use of weather engineering when he sees it, and is steadily reporting and exposing it, see meteorologist Scott Stevens' website http://www.weatherwars.info.

xvii. A notable exception is meteorologist and TV weatherman Scott Stevens, who observes and publishes much information on the weather war going on in the skies over the United States, on his website http://www.weatherwars.info/.

xviii. Particularly professional and hard-hitting coverage of the desperate energy situation is given by Matthew Simmons, head of Simmons & Company International. Highly professional presentations of the real situation in energy worldwide, is available in PDF format from the Simmons website http://www.simmonsco-intl.com/.

xix. T. E. Bearden, "Precursor Engineering: Directly Altering Physical Reality," Explore, 13(2), 2004, p. 50-60. Briefly, there are no forces or force fields in mass-free space, but only in mass systems. For electromagnetic systems, what we call "force" is the ongoing interaction of the force-free EM field in space, with charged mass q. *That ongoing interaction is identically force*. Force is therefore the *ongoing effect* of a more fundamental ongoing causative interaction of curved space with mass.

The field in space is simply an energy pattern, existing as a change to the virtual particle flux of the vacuum (particle physics view) and a curvature of spacetime (general relativity view). It is also a set of Whittaker energy flows, where by "change in energy" we mean "change in the VPF of the vacuum," and "change in the curvature of spacetime". Accordingly, we may formally define energy as spacetime curvature, and also as change in the VPF of the vacuum.

The precursor EM field is simply the EM field as it exists in mass-free space, prior to the field's interaction with matter to form forces and a force field.

The precursor EM field — even a "static" field — is also a set of ongoing Whittaker energy flows that continue so long as the field is merely present. So *unlimited energy flow itself is free for one's taking and using, from any precursor (force-free) field or potential that is established in space. A region of spacetime curvature may be modeled as comprised of a harmonic series of bidirectional longitudinal EM waves —real EM energy flows from the source charges producing the associated EM fields and potentials (EM curvatures of spacetime). One merely has to intercept and collect the freely flowing EM energy from any static field or potential, and prevent any of the collected energy being used in one's circuit to destroy the source of potential or the source of field.* This is easiest accomplished by potentializing the circuit with its charges q pinned, so dq/dt = 0. Then switch away the completely undepleted external source of potential, recomplete the circuit (as with a diode and resistor in series), unpin the electrons and allow current to flow. The new "freely potentialized" circuit will then freely dissipate power in that circuit's loads and losses, until all the initial free potential energy is dissipated. The free energy flow phenomenon and resulting free potential energy change in the absence of current, actually is the physical mechanism that generates the gauge freedom axiom, whereby the potential — and thus the potential energy — of any EM system can be freely changed at will, and without doing work.

Because precursor fields are copious free flows of energy continually extracted from the vacuum via the source charges, one can directly engineer very powerful changes in the energy of the causative force-free precursor field, requiring that one furnish only a bit of work for the switching and structuring and controlling of the freely available energy flows comprising that change. Thus one can almost freely produce a very large precursor energy field set prior to forces and "field work" being involved at all. By subsequently interacting this very large but nearly free precursor energy field set with charged mass in a physical system, extraordinarily large forces and physical effects (work) are then almost freely produced in the mass system. This strongly calls attention to the fact *that there are no "conservation of work" laws in physics!* There are only *conservation of energy* laws. Work is simply changing the form of energy; if the "energy changed in form" is received for free, then so can the resulting work be "for free". In the present scientific view, this is negentropic engineering — and it is perfectly permissible by the laws of nature and of physics.

One must also be careful to distinguish between "internal system work" and external work by the system on something else (an external load). To do internal work, a single joule of EM energy can be "used" (converted in form) again and again, to produce multiple joules of work all internally on the system. *Work* rigorously is the change of form of energy. Given X joules of energy to begin with, one can get X joules of work in one change of form of the energy — and if that work was internal and none of the energy in the new form is lost from the system, one still has all the energy left within the system, just in a different form from what it was in the beginning. If the system still collects this energy in its new form and does not allow its dissipation, it can again be changed in form (e.g., back to the original form) to produce X more joules of work internally to the system. But then the energy is still present, and now in the original form again! Closed looping that "circulation" of changing-form-energy, while preventing any losses of the circulation flow, is a problem for the dogmatic advocate castigating perpetual motion (which is actually Newton's first law of motion). The "perpetual motion while doing continuous work with no additional energy input is impossible" statement contains a false assumption: that each time energy changes form and thereby does work, the work is *external* work by the system on an external load, and the control of the energy in the new form is just "lost" — i.e., the energy in the new form is just dissipated on out of the system, never again to return. *One does not have "dissipation of energy" as work unless the energy in its new form after the work is lost from the system.* If it is not dissipated out of the system, control of it by the system has not been lost. It can again be used to perform additional internal work in the system, though not external work outside the system if it is to continue doing internal work! So it will do that internal system work indefinitely. Note that this form of "perpetual motion and perpetual internal work with no additional energy input" is quite compatible with the conservation of energy law, once one recognizes the difference between internal work and external work, and understands that work is rigorously the change of form of energy, nothing else. It is still quite true that, in the engineering sense where one wishes to do external work on an external load outside the system, then the dissipated energy performing work in the external load is not recovered (usually), and the usual simple thermodynamics emerges.

By precursor engineering (i.e., Russian energetics), one can pay a little bit to almost freely engineer very large energy patterns in mass-free spacetime, prior to their subsequent interactions with matter. Then by simply interacting these "almost freely developed pure energy structures and dynamics" with charged matter, very great work (force effects, or force engines acting in that charged matter) can be almost freely produced, including at great distances. By continuing to propagate a hoary old false EM model that still assumes force fields in space (and thereby still assumes the old material ether), our own scientific leaders are blindly and very determinedly withholding the U.S. scientific community from knowledge and use of precursor engineering (energetics) and therefore from knowledge and use of negative entropy engineering. Since the KGB weapon scientists are using precursor engineering (energetics) in superweapons of great power, the ignorance of our scientific leaders and their refusal to change the terribly flawed CEM/EE model are leading the United States directly into our catastrophic destruction.

xx. Nikola Tesla, "Pioneer Radio Engineer Gives Views on Power," New York Herald Tribune, Sep. 11, 1932.

xxi. E. T. Whittaker, "On the Partial Differential Equations of Mathematical Physics," Mathematische Annalen, Vol. 57, 1903, p. 333-355. Downloadable from www.cheniere.org.

xxii. E. T. Whittaker, "On an Expression of the Electromagnetic Field Due to Electrons by Means of Two Scalar Potential Functions," Proc. Lond. Math. Soc., Series 2, Vol. 1, 1904, p. 367-372. Downloadable from www.cheniere.org.

xxiii. J. D. Jackson, Classical Electrodynamics, 2nd Edn., Wiley, 1975, p. 249.

xxiv. Elsewhere we have formally corrected the present Second Law of thermodynamics, which is an oxymoron implicitly assuming its contradiction has first occurred (to move the system away from equilibrium, thus lowering its entropy — a negative entropy operation). The old Second Law deliberately does not account its assumed *preceding negative entropy operation.* It then accounts only for the entropic decay of the out-of-equilibrium system back to equilibrium — the state of maximum entropy. That is why it prescribes only the production of increasing positive entropy.

Written presently as ($0 \leq \Delta S < +\infty$), the Second Law assumes — but deliberately does not account for — its own contradiction by a prior negative entropy operation ($-\infty < \Delta S \leq 0$), and thus formally is a logical oxymoron and always has been. The old Second Law ($0 \leq \Delta S < +\infty$) must be replaced by the new law ($-\infty < \Delta S < +\infty$), which now also accounts for the previously unaccounted but implicitly assumed prior negative entropy precursor operation.

Very often, this overall new Second Law is applied in systems in two serial phases: first an excitation (potentialization, or asymmetric regauging) phase occurs in which the system departs from equilibrium via the new half-law ($-\infty < \Delta S \leq 0$) The potentialization phase is then followed by the excited system's decay back to equilibrium by the old half-law ($0 \leq \Delta S < +\infty$). For such separate potentialization and decay phases, the second law can be expressed as the equation $(-\infty < \Delta S < +\infty) = (-\infty < \Delta S \leq 0) + (0 \leq \Delta S < +\infty)$. In that form, the old second law is the rightmost term and is easily seen to be only a "half-law". *It merely accounts the decay of a previously excited system, back to equilibrium.* The new overall Second Law is (i) consistent with experiment, (ii) includes the known thermodynamic contradictions to the old Second Law (several are given in Kondepudi and Prigogine, Modern Thermodynamics, Wiley, 1999, p. 459), (iii) includes transient statistical fluctuations in an equilibrium system including their initial negative entropy halves, (iv) removes the logical error in the old law's statement, and (v) now includes negative entropy engineering (precursor engineering, or what the Soviet KGB weapon scientists dubbed "energetics"). Since every charge and dipole in the universe explicitly and freely pours out real observable EM energy at light speed, without any observable energy input (which means with only *virtual-state* energy input), then every charge already demonstrates continuous production of negative entropy and experimentally falsifies the present old Second Law. It does, however, comply with the new and corrected Second Law, which is clearly and experimentally established once one studies the "source charge problem" together with its polarization of the vacuum and its consequent "broken symmetry of opposite charges" solution. Every charge therefore happily and continuously produces negative entropy, of the sort shown theoretically possible by D. J. Evans and Lamberto Rondoni, "Comments on the Entropy of Nonequilibrium Steady States," J. Stat. Phys., 109(3-4), Nov. 2002, p. 895-920. Evans and Rondoni despaired of finding any real system exhibiting such continuous production of negative entropy, but did allow that, for deterministic systems, the "problem remains." After solving the long-vexing source charge problem, we then nominated every charge and dipole (and dipolarity) in the universe as examples of real physical systems continuously producing negative entropy as predicted by Evans and Rondoni.

xxv. Richard P. Feynman, Robert B. Leighton, and Matthew Sands, The Feynman Lectures on Physics, Addison-Wesley, Reading, MA, Vol. 1, 1964, p. 12-2.

xxvi. *Ibid.* Vol. 1, 1964, p. 2-4.

xxvii. See T. E. Bearden, Fact Paper, "Perpetual Motion vs. 'Working Machines Creating Energy from Nothing': With a Discussion of Perpetual Extraction and Emission of Real EM Energy from the Vacuum," available on website www.cheniere.org. The paper also gives a full explanation of the exact reordering and coherent integration mechanism used by every source charge to produce continuous negative entropy.

xxviii. T. E. Bearden, Energy from the Vacuum: Concepts and Principles, Cheniere Press, Santa Barbara, CA, 2002.

xxix. P. Drude, Ann. Physik, Vol. 1, p. 566; Vol. 3, 1900, p. 370, 869.

xxx. See Raymond A. Serway, Physics for Scientists and Engineers, with Modern Physics, Third Edition, updated version, Saunders College Publishing, Philadelphia, 1992, p. 742-745 for a pertinent discussion of drift velocity.

xxxi. Serway, *ibid.* p. 744 shows a typical calculation of a drift velocity of 2.46 × 10exp(-4) m/s.

xxxii. See E. T. Whittaker, A History of the Theories of Aether and Electricity, London, 1958 for a treatise on ether theories, their type, their development, etc.

xxxiii. (a) A. A. Michelson and E. W. Morley, "Influence of motion of the medium on the velocity of light." American Journal of Science, Vol. 31, Series 3, 1886, p. 377-386; (b) — "The relative motion of the earth and the luminiferous aether," American Journal of Science, 4(3), 1887, p. 333; (c) — "On the relative motion of the earth and the luminiferous aether," Phil. Mag. 24(4), 1887, p. 449.

xxxiv. J. D. Jackson, Classical Electrodynamics, 2nd Edn., John Wiley & Sons, New York, 1975, p. 249.

xxxv. Secretary of Defense William Cohen at an April 1997 counterterrorism conference sponsored by former Senator Sam Nunn. Quoted from DoD News Briefing, Secretary of Defense William S. Cohen, Q&A at the Conference on Terrorism, Weapons of Mass Destruction, and U.S. Strategy, University of Georgia, Athens, Apr. 28, 1997.

xxxvi. In any large and powerful organization, essentially the leader does not personally implement anything, but makes decisions and issues orders. The circle immediately surrounding him (first level down in the organization) is his advisors and staff; they also do not personally implement, but relay the orders to appropriate sections of the organization. The second level down is where the large parts (directorates, Sections, etc.) of the organization exist. A manager at that second level down is busy with administration matters, so he personally does not implement. But at the third level down, where the suborganization is now smaller, that manager is usually the implementer. To implement is to perform actions laterally, in and outside the organization, so the third level manager's world is largely a lateral world, with extensive functions, contacts, agreements, etc. with outside groups and agencies. Now call those third level managers and organizations the "implementation entourage" or just the "entourage" for short. This is where all the lateral action happens.

Now apply the bell-shaped normal distribution curve to the entourage managers. At least 9% are lacking in matters such as ethics etc. Shrink that 9% down to 1%, and now one has managers who are in fact very evil persons. They would kill their own grandmother to gain power, control, etc. Law and rules mean nothing except unfortunate encumbrances. Hereafter we will simply call this evil, grasping, conniving 1% "the entourage", simply understanding it is only 1% or so of those managers at that level. But the entourage is the level at which all the skullduggery, murders, assassinations, etc. are planned and either carried out or arranged. We are speaking of every large and powerful organization on Planet Earth! The entourage is dictated by primate dominance, and this big monkey game has been going on in humans as long as there have been humans.

xxxvii. (a) "The Takahashi Magnets and Motor", Infinite Energy, Vol. 1, No. 5&6, Nov 1995/Feb1996, pp. 35-37; (b) Tony Edwards, "'Accident' in lab creates super motor." The [London] Sunday Times, Dec. 10, 1995. Edwards reported an electric motor scooter with top speed of 50 mph and range of more than 500 miles using an electromagnetic engine and magnets developed by Takahashi.

The new scooter, developed by Sciex Corporation of Japan, ran on four ordinary 12-volt car batteries. The magnets were based on powder-metallurgy techniques and special neodymium material which produced a magnet with the world's highest Megagauss Oersted rating at the time. 55 MgOe was the maximum previously, but Takahashi's magnet material achieved 120 MgOe. The Takahashi engine produced 15 horsepower from just a few amperes of electricity. When the scooter was free-wheeling, the engine became a generator and partially recharged the batteries.

Another novel use of the Takahashi magnets was as "battery doublers" to extend the life of rechargeable batteries. Thin inch-wide squares, attached to mobile phone batteries, doubled the amount of charge they retained and allowed the batteries to last twice as long. (c) Also see entries under magnetic Wankel engine, in T. E. Bearden, Energy from the Vacuum: Concepts and Principles, Cheniere Press, Santa Barbara, CA, 2002, p. 363-368.

Any university electrical engineering department worth a hoot can readily build a magnetic Wankel engine, which kills its back mmf and thus continuously develops torque in a single rotary direction. One pays only a tiny bit to kill the back mmf, and self-powering of the system and its load can be achieved by tapping off a bit of power from the shaft-powered generator, and feeding it back to power the switching and control circuit. That none of our EE departments design and build such devices is a sad commentary on the lack of "forward vision" of our scientific community.

xxxviii. (a) Teruo Kawai, "Motive Power Generating Device," U.S. Patent No. 5,436,518. Jul. 25, 1995; (b) see also Bearden , Energy from the Vacuum, 2002, p. 363, 371-375, 500. In 1996 Kawai met with this author and his CTEC board of directors, here in Huntsville, Alabama. After much negotiation on his second trip here, Kawai and CTEC reached agreement. We at CTEC would immediately proceed to set up laboratory and production facilities for producing and marketing worldwide Kawai's self-powering motor, so that in 1997 we would have had such systems in mass production and on the world market everywhere. Our only exclusion would have been Japan, as Kawai had already licensed the marketing rights for that nation. Kawai was backed by the second richest man in Japan, a multibillionaire, and he was financing our effort himself. In short, full financing was already accomplished. The night of the same day that we reached agreement verbally, a private jet arrived post haste from Los Angeles, with Yakuza on board. The next morning Kawai no longer controlled his company, his invention, or his own fate. He was absolutely stunned and shamed, and could hardly lift his head to look me in the eye. He was a totally broken man. Instead, we found ourselves facing the arrogant Yakuza, some with the blue markings and the missing fingers, and very hostile Japanese. The agreement with Kawai was dissolved by the Yakuza right there on the spot. The Japanese team then coldly packed up the Kawai motors that had been shipped to us, and bundled up the Kawai team and left. And that was that. But once one understands the Yakuza's role in this Operations Phase that opened about the first of 2005, one then understands why the Yakuza have suppressed all the Japanese overunity EM systems.

xxxix. Appreciable details on Nathan Stubblefield and his inventions are given by Gerry Vassilatos, Lost Science: New Edition, Adventures Unlimited Press, 2000. Tesla visited Stubblefield and got some of his own impetus from observing the anomalous and awesome electrical power effects achieved by Stubblefield. Many of the stunning effects achieved by Stubblefield have also been successfully replicated by inventor John Bedini.

xl. We give the specific nuclear reaction in T. E. Bearden, Energy from the Vacuum: Concepts and Principles, Cheniere Press, 2002, p. 569, reaction equation (12).

xli. See http://www.pbs.org/wgbh/pages/frontline/shows/cyberwar/interviews/weiss.html.

xlii. According to Italian channel RAI 1, on Sept. 18, 2004 there was an attempt on Putin's life by planting two car bombs along his automobile route. Security forces discovered the cars and dismantled the bombs. Putin has also nullified the election process in Russia, and has thus returned the nation to some of the stronger control mechanisms by the old Communist Party methods. One sees the heavy hand of the Communist faction of the KGB/FGB — which controls the organization and the superweapons and thus controls Russia — as behind the return to greater authoritarian hold over the Russian populace.

xliii. See Matthew Kaminski, "KGB Democracy," Wall Street Journal, Sept. 17, 2004, p. A14.

xliv. For a commentary, see (a) Henry Sokolski, "It's Almost Too Late to Stop Iran," Wall Street Journal, Sept. 27, 2004, p. A18. See also (b) Henry D. Sokolski and Patrick Clawson, Editors, Checking Iran's Nuclear Ambitions, Institute for Strategic Studies, 2004. As of this writing, Iran is continuing on its path to building nuclear reactors capable of furnishing weapons grade materials for nuclear weapons. Further, Russia is again assisting Iran in building the reactors.

xlv. Fatwah, Feb. 1998, issued in the name of the World Islamic Front, and signed by Bin Laden, Zawahiri, and two others.

xlvi. Osama Bin Laden, ABC-TV Interview in Afghanistan, May 1998.

xlvii. The 9/11 Commission Report, Sept. 2004, p. 51-52.

xlviii. Vice President Dick Cheney, October 21, 2001.

xlix. As reported by Niall Ferguson, "Sinking Globalization," Foreign Affairs, 84(2), Mar./Apr. 2005, p. 76-77. Ferguson is Professor of History at Harvard University, a Senior Fellow at the Hoover Institution, Stanford University, and a Senior Research Fellow of Jesus College, University of Oxford.

l. http://www.math.uio.no/~karstent/waves/index_en.html contains a summary article on mathematical attempts to model freak waves and uncover their causes. The article is Kristian B. Dysthe *et al.*, "Freak Waves, Rogue Waves, Extreme Waves and Ocean Wave Climate," June 2003. The authors are all mathematicians at various Norwegian universities and doing research on such waves. As can be seen, these types of waves occur far more frequently than previously suspected, can be larger than previously thought possible, but are also still little understood. Some of the attempted theoretical explanations are summarized in the article, with pros and cons of each being given also.

li. See http://news.bbc.co.uk/1/hi/sci/tech/3917539.stm, "Freak Waves Spotted from Space", BBC News.

lii. Per statistics of the Head Department of Navigation and Oceanography (HDNO) of the Russian Federation Ministry of Defence.

liii. See Bill McGuire, A Guide to the End of the World: Everything You Never Wanted to Know, Oxford University Press, paperback, Aug. 2004 details the phenomenology of large volcano eruptions, supervolcanoes, and tsunamis, among other catastrophic geological events. McGuire is the Benfield Professor of Geohazards and also is the Director of the Benfield Greig Hazard Research Centre, Department of Geological Sciences, University College London, Gower Street, London WC1E 6BT.

liv. See "Supervolcanoes," BBC2, Thursday 3 Feb. 2000; available for downloading from website http://exodus2006.com/transcript.htm.

lv. See http://www.anthro.uiuc.edu/faculty/ambrose. Stanley H. Ambrose is an Associate Professor in the Department of Anthropology and Department of Nutritional Sciences at the University of Illinois Urbana-Champaign, and also is Director of the Anthropology department's Environmental Isotope Paleobiogeochemistry Laboratory.

lvi. E.g., see Jorge A. Vazquez and Mary R. Reid, "Probing the Accumulation History of the Voluminous Toba Magma," Science, Vol. 305, 13 August, 2004, p. 991-994.

lvii. From a transcript of an interview on BBC, available for downloading on website http://www.davidpbrown.co.uk/nota-bene/dna.html.

lviii. See Merritt Ruhlen, The Origin of Language: Tracing the Evolution of the Mother Tongue, Wiley, 1996. There is also persuasive linguistic evidence supporting this bottleneck hypothesis, along the lines presented by Ruhlen. Our special thanks to Marcia Stockton for pointing this out.

lix. E.g., see (a) Stanley H. Ambrose, "Late Pleistocene human population bottlenecks, volcanic winter, and differentiation of modern humans," Journal of Human Evolution, Vol. 35, 1998, p. 115-118; (b) S. H. Ambrose, "Population Bottleneck," in R. Robinson (ed.), Genetics, Macmillan Library Reference, 2002 [a 4-volume encyclopedia of genetics]; (c) M. R. Rampino and S. H. Ambrose, "Volcanic winter in the Garden of Eden: The Toba super-eruption and the Late Pleistocene human population crash," in F. W. McCoy and G. Heiken, (eds.), Volcanic Hazards and Disasters in Human Antiquity, Geological Society of America Special Paper , Vol. 345, 2000, p. 71-82.

lx. For detailed modeling, see Steven N. Ward and Simon Day, "Cumbre Vieja Volcano — Potential collapse and tsunami at La Palma, Canary Islands," American Geophysical Union, 2001. Available from http://www.benfieldhrc.org/SiteRoot/tsunamis/WardandDay.pdf.

lxi. See www.oceansatlas.org/unatlas/issues/emergencies/OceanPhenHDNO/Dangerous.html. The information on hazards to shipping and losses of shipping is provided by the Head Department of Navigation and Oceanography (HDNO) of the Russian Federation Ministry of Defence.

lxii. Marcia Stockton, private communication.

lxiii. James Davison Hunter, Culture Wars: The Struggle to Define America, Basic Books, New York, 1991 (Paperback edition, 1992). Dr. Hunter is Director, Institute for Advanced Studies in Culture, and is also the William R. Kenan Jr. Professor of Sociology and Religious Studies at the University of Virginia.

lxiv. John Zogby, President of the Zogby International Survey firm. Quoted from John Mariani, "Zogby Describes Divided Nation," The Post Standard, Sept. 24, 2004. Article is carried on website http://www.syracuse.com/elections/poststandard/index.ssf?/base/news-1/1096017657151770.xml

lxv. Details of these dramatic deaths, and how they were technically achieved, are given in T. E. Bearden, Fer-de-Lance, Cheniere Press, updated 2003.

lxvi. Stanislav Lunev (with Ira Winkler), Through the Eyes of the Enemy, Regnery, Washington, 1998. Lunev is the highest ranking GRU defector. He is a former Colonel in the GRU, the military counterpart of the KGB. In p. 22-33, he summarizes the Spetznaz capabilities. On p. 22-27, he summarizes Spetznaz use of nuclear weapons already on American soil. On p. 26, he gives some of the ways in which the Soviets easily brought nuclear weapons into the U.S. On p. 30 he confirms Russian seismic weapons. Use of EMP weapons is on p. 30-31.

lxvii. U.S. General George S. Patton Jr., in his speech to Third U.S. Army troops on the eve of Operation Overlord in 1944.

lxviii. As a typical example, see Colin S. Gray, "Thinking Asymmetrically in Times of Terror," Parameters, U.S. Army War College Quarterly, Spring 2002, p. 5-14. The basic message of the article is that asymmetric war is really nothing new, even the concept doesn't have any real content, and the terrorists can do a little damage and we will take some damage, but we will survive as always. In the vernacular, it proposes that we shouldn't sweat the small stuff; everything is still AOK and there isn't any really new or unique problem. Now contrast this with the recent registration of a Yakuza scalar interferometer on the Yellowstone caldera, so that in a single shot most of the United States can be destroyed. We really do not recall seeing that kind of thing occur in our past "asymmetric" wars, nor do we see initiation of earthquakes such as the 9.3 quake on the seafloor on Dec. 26, 2004 that initiated the catastrophic tsunamis in Asia. But one notes that, in 2002 — five years after Defense Secretary's confirmation of special terrorist EM weapons generating earthquakes, evoking volcanoes into eruption, and controlling our climate and weather — the official publication Parameters of the U.S. Army War College apparently has no knowledge whatsoever of such terrorist capabilities *or of such terrorist demonstrations.*

lxix. Missing and unaccounted, in the official Ukraine records and accounting for its nuclear weapons transfer to Russia. The reader should understand that there are no *officially acknowledged* missing nuclear weapons, and there likely never will be! In the real world, there turns out to be hundreds of such weapons that have "mysteriously disappeared from accounting," and then in the hue and cry and charges and counter charges, the hullabaloo has just faded away and been ignored.

lxx. The Ministry of Defense reported on transfer and destruction by the Ukraine of 3772 nuclear warheads, but the general staff reported there were more than 4,000 to begin with. In other words, something other than transfer or destruction was done with some 230 or so Ukrainian nuclear warheads.

lxxi. Even under U.S. agreement with the Russians (START II) for dismantling some of the nuclear weapons, there has been no truly accurate accounting. E.g., quoting from Jo L. Husbands, BITS Research Report 97.1, Part II: Domestic and International Politics: The U.S. Response to the Russian Nuclear Legacy, Nov. 1997: *"From the outset, the Russian government was unwilling to permit the United States to have a role in the actual dismantlement of former Soviet/Russian nuclear warheads. The United States therefore cannot say with certainty how many warheads have actually been dismantled at the four Russian dismantlement facilities..."* In other words, much of it is smoke and mirrors.

So the reports of precisely how many warheads were transferred and/or dismantled and destroyed are not accurate at all in their "total numbers," and they never have been. Instead, the infamous and deceitful catch-all phrase so long used by our intelligence agencies to cover ignorance was invoked. The phrase is *"We have no evidence that...,"* which usually is a bald-faced attempt to cover up something which we absolutely do not know the truth or falsity of, or that is embarrassing, or that is a political hot potato. Usually the evidence is indeed there somewhere, but it indicates something contrary to the prescribed "official intelligence position" of the agency. Analysts who push too hard in advancing what the intelligence database really states, when it is against the official policy position, do not fare well or last very long in the U.S. intelligence community.

lxxii. In 1996 Gen. Alexander Lebed, then President Boris Yeltsin's national security adviser, reported that 84 suitcase-size nuclear weapons were missing from Russia's arsenal. Other Russian officials denied the claim, and U.S. officials discounted it. The matter was never resolved, but just gradually ignored.

lxxiii. Paul L. Williams, Al Qaeda: Brotherhood of Terror, Alpha Books, Indianapolis, 2002 states Russian and U.S. intelligence sources are aware that bin Laden purchased 20 suitcase nuclear weapons in 1998 from Chechen Mafia agents for $30 million in cash and a quantity of heroin whose street value was $700 million. Williams is a former FBI consultant and the author of several important books on terrorism.

lxxiv. Eliot Marshall; David Malakoff, and Constance Holden, "Bioweapons Cleanup," Science, Vol. 295, 25. Jan. 2002, p. 603.

lxxv. Simon Reeve, The New Jackals: Ramzi Yousef, Osama bin Laden, and the Future of Terrorism, Northeastern University Press, Chicago, 1999, p. 208.

lxxvi. The study reference is: U.S. Congress, Office of Technology Assessment, Proliferation of Weapons of Mass Destruction: Assessing the Risks, Government Printing Office, Washington, 1993, p. 54.

lxxvii. Richard K. Betts, "The New Threat of Mass Destruction," Foreign Affairs, 77(1). Jan/Feb. 1998, p. 32.

lxxviii. Manitoba Health, Emergency Treatment Guidelines: General – Triage, Aug. 2004. This Canadian guide is deliberately referenced to show that the triage procedure is a necessary medical emergency procedure, recognized internationally.

lxxix. Our own National Institutes of Health (NIH) knows nothing about higher group symmetry electrodynamics, knows nothing about precursor engineering (the Soviet energetics science), at best knows the standard but seriously flawed electrical engineering and classical Maxwell-Heaviside electrodynamics, never understood the induction of health changes and diseases in U.S. Embassy personnel in Moscow by the decades of so-called "microwave radiation of the U.S. Embassy," knows nothing of the mechanism proved by Prioré and important French scientists to yield revolutionary electromagnetic cures for dread terminal illnesses (such as cancer, clogging of the arteries, some infectious diseases tested, etc.), knows absolutely nothing at all about Soviet bioenergetics or its mechanisms and capabilities, and certainly knows nothing of (and does not even "believe") the startling capabilities of Soviet/Russian/FSB/KGB psychoenergetics and its biogenesis ability to build an actual living mind and mind operations in the laboratory, then clone or reproduce it or them by the millions.

NIH and its subsidiary organizations seem interested in maintaining the status quo in medical science and adhering scrupulously to present medical treatment mechanisms and theoretical understanding. They seem to block and deny any and all more advanced and innovative research, particularly if advanced group symmetry electrodynamics is involved.

They have little knowledge of the actual healing mechanism of the body (its cellular regeneration system, which is a highly special electromagnetic system), and do not understand the higher group symmetry electrodynamics of that system. Further, they do not intend to even study the kind of "novel" electrodynamics and electrodynamics' mechanisms used by the body for all its cellular healing processes. Instead, they are absolutely determined to continue with drugs, vaccinations, surgery, antibiotics, and conventional medical therapies. They therefore are determined that the majority of those coming mass casualties from anthrax spray attack etc. shall be untreatable, so that most of those poor patients will be dragged aside and left to die without treatment, under Triage procedures. Hitler in all his death camps may not have produced as many millions of deaths as will this dogmatic policy of our medical science institutions.

Consequently, we are 50 years behind where we should be in medical therapy technology. We are also extremely vulnerable to mass casualties generated in our population centers by terrorist WMD attacks. Whether intentional or unintentional, NIH's actions and policies are designed to keep us in that situation indefinitely.

lxxx. (a) Vlail Kaznacheyev and L. P. Mikhailova, Ultraweak Radiation in Intercellular Interactions, [in Russian], Novosibirsk, 1981. An English exposition of much of the Kaznacheyev work is given in (b) Vlail Kaznacheyev, "Electromagnetic Bioinformation in Intercellular Interactions," Psi Research, 1(1), Mar. 1982, p. 47-76. See also (c) Vlail Kaznacheyev, "Apparent information transfer between two groups of cells," Psychoenergetic Systems, 1(1), Dec. 1974, p. 37. (d) Vlail Kaznacheyev *et al.*, "Distant intercellular interactions in a system of two tissue cultures," Psychoenergetic Systems, 1(3), Mar. 1976, p. 141-142. In the same issue, see (e) "A Comment" by William A. McGarey, p. 143; and (f) "A Comment" by Arthur C. Hastings, p. 143-144.

lxxxi. Aitchison, I. J. R., "Nothing's Plenty: The Vacuum in Modern Quantum Field Theory," Contemporary Physics, 26(4), 1985, p. 372.

lxxxii. See Rick Weiss, "NIH to Set Stiff Restrictions on Outside Consulting," Washington Post, Aug. 4, 2004, p. A01.

lxxxiii. "US NIH Wants 1 Year Moratorium on Drug Co. Collaborations," Dow Jones Newswires, 9-24-04 1318ET.

lxxxiv. (a) The U.S. Patent is "Pathogenic Mycoplasma", US Patent No. 5,242,820, issued September 7, 1993. Dr Lo is listed as the Inventor and the American Registry of Pathology, Washington, DC, is listed as the "Assignee"; (b) for a summary, see Donald W. Scott, "Mycoplasma: The Linking Pathogen in Neurosystemic Diseases," Nexus Magazine, Aug. 2001. Particularly see (c) US Senate, Ninety-fifth Congress, Hearings before the Subcommittee on Health and Scientific Research of the Committee on Human Resources, Biological Testing Involving Human Subjects by the Department of Defense, 1977; released as US Army Activities in the US Biological Warfare Programs, Volumes One and Two, 24 February 1977. See also (d) Donald W. Scott and William L. C. Scott, The Brucellosis Triangle, The Chelmsford Publishers, Sudbury, Ontario ,Canada, 1998.

lxxxv. The interested reader is referred to our "Porthole Briefing" on website www.cheniere.org.

lxxxvi. (a) T. E. Bearden, Energetics: Extensions to Physics and Advanced Technology for Medical and Military Applications, Mar. 21, 1998, 200+ page enclosure to CTEC Letter to Gen. (Ret.) Walter Busby, Deputy Secretary of Defense for Counterproliferation and Chemical and Biological Defense, March 21, 1998., also the enclosure to CTEC letter to Major General Thomas H. Neary, Director of Nuclear and Counterproliferation, Office of the Deputy Chief of Staff, Air and Space Operations, HQ USAF, May 4, 1998. Copies were also sent as enclosures to letters to NIH, CDC, and several other government agencies and laboratories. (b) T. E. Bearden, "Method, System and Apparatus for Conditioning Electromagnetic Potentials, Fields, and Waves to Treat and Alter Matter," Application No. 60/339,122, Dec. 7, 2001. This is the provisional patent application that the present author openly released to the public domain in Dec. 2002, and it is posted on www.cheniere.org. (c) See also T. E. Bearden, Porthole Briefing, available on www.cheniere.org.

lxxxvii. (a) B. Baker; J. A. Spadero, and R. O. Becker, "Electrical stimulation of articular cartilage," Annals of New York Academy of Sciences, Vol. 238, 1974, p. 491-499; (b) R. O. Becker, Carlton F. Hazlewood, Abraham R. Liboff, and Jan Walleczek, "Electromagnetic Applications In Medicine," NIH-OAM Electromagnetics Panel Report, Jan. 15, 1993; (c) Robert O. Becker and Andrew A. Marino, Electromagnetism and Life, State University of New York Press, Albany, 1982; (d) Robert O. Becker, "The neural semiconduction control system and its interaction with applied electrical current and magnetic fields," Proceedings of the XI International Congress of Radiology, Vol. 105, 1966, p. 1753-1759, Excerpta Medica Foundation; Amsterdam; (e) R. O. Becker, "The direct current field: A primitive control and communication system related to growth processes," Proceedings of the. XVI International. Congress of Zoology, Washington, D.C., Vol. 3, 1963, p. 179-183; (f) R. O. Becker, Charles H. Bachman, and Howard Friedman, "The direct current system: A link between the environment and the organism," New York State Journal of Medicine, Vol. 62, April 15, 1962, p. 1169-1176.

lxxxviii. R. O. Becker and David G. Murray, "The electrical control system regulating fracture healing in amphibians," Clinical Orthopaedics and Related Research, No. 73, Nov.-Dec. 1970, p. 169-198. This is the definitive technical exposition by Becker of his electrical control system ideas and findings.

lxxxix. (a) R. O. Becker and D. G. Murray, "A method for producing cellular dedifferentiation by means of very small electrical currents," Trans. N.Y. Acad. Sci., Vol. 29, 1967, p. 606-615; (b) Becker, R. O., and Joseph A. Spadaro, "Electrical stimulation of partial limb regeneration in mammals," Bulletin of the New York Academy of Medicine, Second Series, 48(4), May 1972,
p. 627-64' (c) C. A. L. Bassett; R. O. Becker, and R. J. Pawluk, "Effects of electric currents on bone in vivo," Nature, Vol. 204, Nov. 14, 1964, p. 652-654.

xc. T. E. Bearden, Porthole Briefing, available on www.cheniere.org.

xci. See (a) A. J. Berteaud and A. M. Bottreau, "Analyse des rayonnements électromagnétiques émis par l'appareil Prioré," [Analysis of the electromagnetic radiations emitted by the Prioré apparatus], D.R.M.E., 1971, p. 3-12; (b) A. J. Berteaud, A. M. Bottreau, A. Prioré, A. N. Pautrizel, F. Berlureau, and R. Pautrizel, "Essai de corrélation entre l'évolution d'une affection par Trypanosoma equiperdum et l'action d'une onde électromagnétique pulsée et modulée," [Trial of the correlation between the evolution of a disease by Trypanosoma equiperdum and the action of a pulsating and modulated electromagnetic wave], Compt. Rend. Acad. Sci. (Paris), Vol. 272, 1971, p. 1003-1006; (c) R. Cambar, "Rapport general des travaux de la Commission de Contrôle constituée en vue de vérifier l'un des effets biologiques obtenu par l'utilisation de l'appareillage de Prioré A. Bordeaux," [General findings of the work of the control commission formed to verify one of the biological effects obtained by use of the apparatus of A. Prioré at Bordeaux], 1969, 1 vol.; (d) R. Courrier, "Exposé par M. le Professeur R. Courrier, Secretaire Perpetuel de L'Academie des Sciences fait au cours d'une reunion a L'Institut sur les effets de la Machine de M.A. Prioré le 26 Avril 1977," [Presentation by Professeur R. Courrier, Perpetual Secretary of the Academy of Sciences, made at the meeting of the Academy on the effects of the machine of M.A. Prioré.]; (e) G. Delmon and J. Biraben, "La croissance du carcinome de Guerin sour l'action de champs électromagnétiques," [The growth of carcinoma treated by the action of electromagnetic fields], Rev. Path. Comp., 3(2), 1966, p. 85-88; (f) G. Mayer, A. Prioré, G. Mayer and R. Pautrizel, "Action de champs magnétiques associés à des ondes électromagnétiques sur l'orchite trypanosomienne due lapin," [Action of magnetic fields associated with electromagnetic waves on the typanosomian orchitis of the rabbit], Compt. Rend. Acad. Sci. (Paris), Vol. 274, 1972, p. 3011-3014; (g) Prioré's doctoral thesis (which was rejected when the project was suppressed) is: Guérison de la Trypanosomiase Expérimentale Aiguë et Chronique par L'action Combinée de Champs Magnétiques et D'Ondes Electromagnétiques Modulés [Healing of intense and chronic experimental trypanosomiasis by the combined action of magnetic fields and modulated electromagnetic waves], thesis submitted in candidacy for the doctoral degree, 1973; (h) Eric Perisse, Effets des Ondes Electromagnetiques et des Champs Magnetiques sur le Cancer et la Trypanosomiase Experimentale [Effects of Electromagnetic Waves and Magnetic Fields on Cancer and Experimental Trypanosomias], Doctoral thesis, University of Bordeaux No. 83, March 16, 1984; (i) For a limited background and light technical coverage of Prioré's work for the educated layman, see T. E. Bearden, "Vacuum Engines and Prioré's Methodology: The True Science of Energy-Medicine, Parts I and II," Explore!, 6(1), 1995, p. 66-76; 6(2), 1995, p. 50-62; (j) For somewhat more specific technical details, see T. E. Bearden, Energetics of Free Energy Systems and Vacuum Engine Therapies, Tara Publishing, Internet node www.tarapublishing.com/books, July 1997; (k) for a summary of the entire Prioré affair, see Christopher Bird, "The Case of Antoine Prioré and His Therapeutic Machine: A Scandal in the Politics of Science," Explore!, 5(5-6), 1994, p. 97-110; (l) For detailed coverage of the entire Prioré affair, see Jean-Michel Graille, Dossier Prioré: Une Nouvelle Affaire Pasteur, [The Prioré Dossier: A New Pasteur Affair?], De Noel, Paris, 1984 [in French]; (m) For a technical description of the ordinary EM and physics characteristics of the Prioré radiation (but not the active mechanism), see A. J. Berteaud and A. M. Bottreau, "Analyse des rayonnements électromagnétiques émis par l'appareil Prioré," ["Analysis of the electromagnetic radiations emitted by the Prioré apparatus"], D.R.M.E., 1971, p. 3-12. Prioré's patents are (n) Antoine Prioré, "Apparatus for producing radiations penetrating living cells," U.S. Patent No. 3,368,155, Feb. 6, 1968; (o) Antoine Prioré, "Method of producing radiations for penetrating living cells," U.S. Patent No. 3,280,816, Oct. 25, 1966, (p) Antoine Prioré, "Procede et dispositif de production de rayonnements utilisables notamment pour le traitement de cellules vivantes," [Procedure and Assemblage for Production of Radiation Especially Serviceable for the Treatment of Living Cells], Republique Francais Brevet d'Invention P.V. No. 899.414, No. 1,342,772, Oct. 7, 1963.

---

xcii. See Laurie Garrett, "The Nightmare of Bioterrorism," Foreign Affairs, 80(1), Jan. /Feb. 2001, p. 76-89. The quote is from p. 77.

xciii. As reported in "Alert as Nuclear Bombs Go Missing," The Drudge Report, Sept. 17, 2002.

xciv. Matthew Bunn and Anthony Wier, "Preventing a Nuclear 9/11," Washington Post, Sept. 12, 2004, p. B07.

xcv. *Ibid.*

xcvi. Paul L. Williams, Osama's Revenge: The Next 9/11: What the Media and the Government Haven't Told You," Prometheus Books, 2004. Williams is a former FBI consultant and author of several treatises on terrorism.

xcvii. Paul L. Williams, Al Qaeda: Brotherhood of Terror, 2001.

xcviii. See (a) T. E. Bearden, Fer-de-Lance, Cheniere Press, 2003 (updated) for some details on quantum potential weapons and their usage. Also see (b) T. E. Bearden, Gravitobiology, Cheniere Press, 2003 (updated) for additional details. Some five nations of the world now possess QP weapons, and a sixth nation is obtaining them. China, Russia, Brazil, Israel, and one other friendly nation are the five present nations.

xcix. For background on such tsunamis when naturally induced, see David R. Tappin, Editor, Submarine-slump-generated Tsunamis, Special edition of Marine Geology, Volume 203, Issues 3-4, 30 January 2004, pages 199-386.

c. Paul H. B. Shin, "Killer wave lying in wait: Volcano could unleash horror," New York Daily News, Aug. 29, 2004, http://www.nydailynews.com/news/story/226886p-194881c.html.

Other best-sellers by Tom Bearden

Books

Energy From The Vacuum: Concepts and Principles

Fer-De-Lance: Briefing on Soviet Scalar Electromagnetic Weapons

Excalibur Briefing: Explaining Paranormal Phenomena

AIDS: Biological Warfare

Gravitobiology: A New Biophysics

The Final Secret of Free Energy

Videos

Floyd Sweet Secrets

Tesla's Secret and the Soviet Tesla Weapons

AIDS as Biological Warfare: A Possible Electromagnetic Solution

Secret Superweapons that Drive Disarmament Negotiations

Soviet Weather Engineering over North America

Tom Talks Tesla

All Tom Bearden titles may be purchased at www.cheniere.org.